The Art of Recording

The Creative Resources of Music Production and Audio

William Moylan

VAN NOSTRAND REINHOLD
New York

An AVI Book
(AVI is an imprint of Van Nostrand Reinhold)

Library of Congress Catalog Card Number 91-23644
ISBN 0-442-00669-1

Printed in the United States of America

Van Nostrand Reinhold
115 Fifth Avenue
New York, New York 10003

Chapman and Hall
2-6 Boundary Row
London, SE1 8HN, England

Thomas Nelson Australia
102 Dodds Street
South Melbourne 3205
Victoria, Australia

Nelson Canada
1120 Birchmount Road
Scarborough, Ontario M1K 5G4, Canada

16 15 14 13 12 11 10 9 8 7 6 5 4 3 2 1

Library of Congress Cataloging-in-Publication Data

Moylan, William.
The art of recording: the creative resources of music production and audio / William Moylan.
p. cm.
Includes bibliographical references and index.
ISBN 0-442-00669-1
1. Sound recording industry. 2. Music trade. 3. Sound recordings—Production and direction. I. Title.
ML3790.M63 1992
781.49—dc20 91-23644
CIP

Contents

Preface

Recording music is a creative process, involving certain artistic decisions. The final shape of a music recording, and possibly of the music itself, is determined by the recording medium and how it is used. The elements that artistically shape the artistry of the music recording are also used in all other areas of audio-recording production.

The concept of recording as an art is widely accepted. Just what makes recording an art has not been well defined. *The Art of Recording* defines the artistic aspects of audio recording and looks at the broad and fundamental concepts that make music and audio recordings art. By addressing these global issues, it defines the dimensions and elements of the art, applies those elements to the recording process, and presents a method for evaluating music recordings. It presents an in-depth study of the artistic aspects of recorded sound, including music recordings.

To provide an adequate understanding of the discipline of recording music, this book should be supplemented by studies in recording techniques, recording technology, and audio theory; sound synthesis and electronic/computer music; acoustics and psychoacoustics; music composition and music theory; mathematics, physics, engineering, and computer science; and poetry and language.

People hoping to enter the field of music/audio recording, will need to be prepared in these areas–some areas more–than others depending on the positions the individual is seeking. These important, related topics will not be covered in detail in this book. Instead, the relationships of these topics to the artistic resources of audio recording will be explored.

This book has been written to be easily understood by people with limited formal backgrounds in acoustics, engineering, physics, math, and music.

It is assumed that the reader is either a professional in some aspect of the audio recording industry, a student preparing for the industry, or an interested amateur (with a background similar to one of the previous two categories).

This book should be used: (1) as a resource book for all people involved in audio production; (2) as a textbook for audio production-related courses and/or listening skills-related courses, in sound recording technology (music engineering), media/communications, or music composition programs; or (3) as a self-learning guide, for the motivated student or interested amateur.

The portions of the book addressing sound quality evaluation will be directly applicable to all individuals who work with sound. Audio engineers in technical areas will benefit from this knowledge and the related listening skills, as much as individuals in creative areas. All those people who talk "about" sound can use the approach to evaluating sound presented herein, and use this book as a resource and self-learning guide.

This book will be useful to the following list of professionals and to those people who are studying the discipline, in hopes of entering one of these fields:

- Music recording engineers and producers
- Media recording engineers and producers
- Engineers and producers of live concert sound
- Engineers and producers of theater sound
- Broadcast audio-recording engineers and producers
- Audio for visual recording engineers and producers
- Film sound mixers
- Special effects and foley recording engineers and producers
- Sound designers
- Composers and arrangers of electronic or computer music
- Composers and arrangers of film/television/video music
- Composers and arrangers of music for recordings

People occupying positions in these areas will be collectively called "recordists," throughout the book.

As a resource work for those in the industry, *The Art of Recording* is designed to contribute to the professional development of recordists. The author seeks to provide a clearly defined approach to the creative aspects of recording and hopes to expand the current professionals' creative potential, skills in critical and analytical listening, and sensitivity to accurate and meaningful communication about sound quality.

Many people actively engaged in the creative and artistic roles of the industry are hard-pressed to describe their creative thought processes. The actual materials they are crafting have not previously been well defined. Current professionals have probably had little guidance in identifying the skills required in audio-recording production; it is likely that they developed their current skills intuitively, with little outside assistance. A recordist may already have highly developed creative abilities and listening skills, yet be unaware of the dimensions of those skills. This book will address these areas and will assist current professionals in discovering new dimensions within their personal, creative voice.

No existing textbooks discuss the creative aspects of recording music or audio. No textbooks discuss and develop the listening skills required to evaluate recordings for technical quality and/or for their artistic treatment of the recording medium. *The Art of Recording* may be used as a textbook in all these areas, from beginning through the most advanced levels.

The book may be used in a wide variety of courses, in many college and university degree programs, and in vocational-type programs in audio and music recording. Music engineering or music production (sound recording technology) programs; communications, media, film, radio/television, or telecommunications programs; and music composition programs emphasizing electronic/computer music composition; will all have courses that speak to the artistic aspects of recording music (and sound).

The book is well suited to developing the student's music-production techniques and thought processes. The artistic dimensions of recording production are defined and applied to the recording process. This book is also designed to stimulate thought about the recording process being a collection of creative resources. The resources are then applied to the act of creating a music performance (making a recording).

As a textbook for sound evaluation and listening skill development, *The Art of Recording* can be used at various levels of competency. The instructor may determine the level of proficiency required of the students, in their performance of listening evaluations. The book presents a method for evaluating sound and applies the method to a number of commercially available recordings. The development of listening skills is a lengthy and involved process, that will go through many stages of accomplishment. The book can be used by students at the beginning of their academic studies or at the most advanced levels, by graduate students as well as undergraduates.

Graphing the activity of the various artistic elements is important for developing aural and evaluation skills, especially during beginning studies. It is also valuable for performing in-depth evaluations of recordings, thus allowing the developing students to study how the artistic aspects of recording have been used by accomplished recording producers.

This process of graphing the activity of the various artistic elements is also a useful documentation tool. Working professionals through beginning students will find the process useful.

It is hoped that this book will help create a new vocabulary. Music and audio recording require communication "about" sound, between the many artistic and technical people of the industry. We in the industry presently function without this meaningful exchange of ideas; we often do not communicate well.

In order for communication to occur, a vocabulary must be present. The vocabulary should contain terms that address sound experiences that are common between all humans. Terms must mean the same thing to the people involved and must apply to something specific within the sound. When terminology is related to the subjective interpretations of one of the listeners, accurate communication will not occur; the individuals attempting to communicate will not have the same set of shared experiences. A wide-spread acceptance of an accurate and meaningful vocabulary would benefit our industry. This book may serve as a meaningful point of departure for an "audio recording vocabulary" to be devised.

The Art of Recording is the product of my experiences as an educator in sound recording technology (music and technology), of my thought processes and observations as a composer of acoustic music, as well as music for recordings (electronic and computer music, and recording productions), of my creative efforts as a recording producer, and of my attempts to be transparent as a recording engineer. It is also a product of my research into how people hear music and sound, as reproduced through loudspeakers (aural perception, music cognition), and into using the recording medium to enhance artistic expression, especially in music.

This book has evolved, to its present form, from my doctoral dissertation (*An Analytical System for Electronic Music*), through eight years of instructional materials in my courses at the University of Massachusetts–Lowell (the former University of Lowell) as well as teaching recording production techniques at the Aspen Audio-Recording Institute, through a series of papers presented to the Audio Engineering Society (1985, 1986, and 1987), and through research into how the recording medium has been utilized to add new, artistic dimensions to music recordings. This book, and the concepts and methodology it presents, have gone through many stages of development since 1981.

ACKNOWLEDGMENTS

A number of people have directly contributed to shaping some of the fundamental concepts of this book.

Dr. Cleve L. Scott, my friend and dissertation advisor, presented me with the perspective that, beyond our conceptions of music and technology, lies the art of music.

Dr. Jean Eichelberger Ivey, my first composition teacher, encouraged me to conceptualize the electronic music medium as a chamber ensemble (equivalent to a string quartet).

Pierre Schaeffer (pioneer of *musique concrète* and of sound quality evaluation) kindly engaged in correspondence with me from 1982 to 1983; his insights led me in directions of thought and research that largely defined my ideas on the dimensions of the artistic elements, as well as my approaches to their evaluation.

My students in the Sound Recording Technology program at the University of Lowell (now the University of Massachusetts–Lowell) have worked through the concepts of this book, in my Aural Perception of Timbre and Space and Recording Technology I courses (both courses are required of all SRT majors). The classes of 1985 and 1986 worked through many of the initial concepts and materials of the book; the classes of 1987 through 1990 worked through various exercises and versions of the concepts as they were developed and refined; and the classes of 1991 and 1992 worked with the materials of the book, in their present state. Successes, failures, frustrations, and feelings of accomplishment, created by understanding the intricacies of broad concepts, were all experienced by these students. Their experiences served as meaningful gauges during the book's formation. My special thanks to these serious and gifted young people for their (largely unknowing) contributions to this project.

Thanks to my friends and colleagues who reviewed various portions and versions of this manuscript, and offered suggestions: Willis Traphagan, Scott Kent, Fred Malouf, Tomasz Letowski, and Wayne Kirby.

Introduction

What makes recording music an art?

What makes music recording a unique medium for artistic expression?

What is different between a music recording and a live music performance?

Why is the recordist (recording producer or engineer) considered an artist?

How does the recording process shape a piece of music?

It is widely recognized that the recording process shapes music. Changes in sound quality are caused by the recording process. They occur under the control of an individual that shapes the music recording.

Changes in sound quality that are created by recording do not occur in nature. They are unique to audio recordings, and give recorded music (or music reproduced over loudspeakers) a set of unique sound characteristics. They are perceived as part of the experience of listening to recorded and reproduced sound.

The unique sound qualities of recording contribute to the character of a piece of music; they become part of the musical composition. The available resources for shaping a piece of music have thus been extended to include the unique sound qualities of music recordings. The person controlling or creating these sound qualities (the recordist) is functioning as a creative artist, as a musician.

OVERVIEW OF ORGANIZATION AND MATERIALS

The above questions will be answered during the course of *The Art of Recording*. The artistic dimensions of recorded music, necessary to answer the above questions, have not previously been defined. This book will devise definitions and will demonstrate how the dimensions are shaped in the recording process. The book will then present a method for evaluating how the dimensions are used in music recordings and for allowing the reader to determine creative applications for the dimensions.

Accordingly, the book is divided into three equal parts:

- Part 1–defines the artistic elements
- Part 2–applies the artistic elements to the recording process
- Part 3–evaluates how the artistic elements have been commonly used in production practice

Part 1

Part 1 is divided into three chapters. To begin defining the artistic dimensions of music recordings, sound must be understood. The states of sound in air, in human perception, and as applied to music are followed in the sequence of understanding the meaning of sounds. The processes that occur in moving from one state of sound to another are emphasized. The distortions that occur in the transfer processes are recognized and evaluated.

Sound as a resource for artistic expression is the basis for Chapter 2. It encompasses the unique sound qualities of recorded sound, and their potentials for artistic expression. This is followed by an examination of the musical message itself, and how musical materials are perceived by the listener, for the understanding of musical messages and communication.

Part 1 centers around the listening process itself. It brings to light the importance of the listener in the communication of musical materials.

People listen to sound and music at various levels of intellectual involvement. They perceive more detail in the music as they move from a passive, undirected "journey" through the emotive states of the music, through an understanding of any literary or extra-musical conceptions in the music (with the presence and influences of any text or other image-inducing associations), through the aesthetic listening experience (appreciating the interrelationships of the abstract musical materials).

The levels of detail of the aesthetic listening experience allows analytical listening. When listening analytically, the listener is actively engaged in seeking knowledge about the characteristics and interrelationships of the musical materials.

The listener's subjective impressions about the music are a central concern only in the most passive of the listening levels. As the attention of the listener is drawn to greater levels of detail, the listener is taken from being moved by the sensuous aspects of music towards an intellectual fascination with the materials of the music and their interrelationships–to the aesthetic or the artistic message(s) or meaning of the work.

Listening for information is a more sophisticated process than simply experiencing the emotive aspects of the music. This is not a qualitative judgement; neither approach is necessarily "better," or perhaps more rewarding, for the particular individual, during recreational listening to music. When the listening process is part of the professional recordist's responsibilities, it becomes a different matter. How one listens, and what one listens for, is a central concern for the recordist. Listening is one of the primary responsibilities of the recordist.

At the greatest level of detail, and requiring the most active listening process, are *analytical listening* and *critical listening*. The professional recordist will have an ability to listen for specific information in these ways; the young person who wishes to enter the recording industry must develop these listening skills.

Critical listening is concerned with the process of evaluating the most minute aspects of the sound itself, out of time and context. The sound is momentarily disassociated from the context of the music and is available to be evaluated for its unique characteristics. Critical listening seeks to define and describe sound in objective terms, not allowing the listener's impressions of the sound to enter into the evaluation process.

Analytical listening seeks the same detailed information about sound quality. It is listening within the context of the piece of music and relating the sound materials to one another and to the music as a whole. Analytical listening also seeks to evaluate the listening experience as objective information; subjective impressions about the music is not a central element in this way of listening.

Part 2

Part 2 applies the artistic elements to the recording production process. It will present the concepts and thought processes of recording production as they are related to the artistic aspects of recording and the artistic elements of sound. The recordist will work through these concepts during the creative processes of making a music recording.

Part 2 will define general principles. It will not present specific models of creative approaches to the medium, nor will it address specific pieces of equipment or specific recording techniques. The principles covered will allow the reader to conceptualize music recording as a creative process; it will place the reader in the position of guiding the artistic product from its beginning as an idea, through its development during the many stages of the recording sequence, to its final form.

This approach will systematically explore the creative potentials of the audio recording medium.

Two possible recording production sequences are presented and contrasted. Through these two scenarios, the artistic concerns of the recording production process are evaluated. The resources of the recording process are considered for their potential to capture sound qualities, to perform the music recording, and to generate the relationships of a piece of music. The artistic roles (or functions) of the recordist are contrasted in relation to the sequences.

The recordist will learn to explore the creative potentials of the medium's tools (equipment) as musical instruments and will develop an artistic sensitivity through the study of the creative works of others (Part 3).

Applying the artistic elements of recording to the recording process (Part 2) and evaluating the artistic elements in the recording process (Part 3) will often occur simultaneously in production practice.

They are, however, two distinct processes. They are presented separately for clarity and for a thorough presentation. The two are interdependent, when considering the evaluation of sound that occurs during the recording production process.

Part 3

In total, part 3 presents a complete method for evaluating the dimensions of sound in music and audio recordings. The need for sound evaluation and the contexts for sound evaluation in music recordings are established.

Each element of sound is evaluated separately. The creative resources of each element are presented. A method of evaluation has been specifically devised for each individual element (as per their potentials in music recordings). The method of evaluation for the individual artistic elements are accomplished in relation to a complete, interrelated system of evaluating sound in music recordings.

The method will also seek to develop the listening skills of the reader and to provide the basis for meaningful and accurate communication on sound content and quality.

The method will progress from simple concepts and listening processes to the most complex. The method will build on experiences that are most easily learned and will evolve systematically to the most difficult. Listening experiences that many audio professionals or intermediate-level musicians may have already acquired (at least intuitively) are incorporated into the beginning listening exercises of the method.

The reader will eventually gain the experience and knowledge that will allow them to perform quite complex listening and analytical tasks. These complex listening evaluations are commonly used by professionals in the audio recording industry. People who are not equipped with the knowledge and skills to perform these listening evaluations accurately, or to understand or identify what they are hearing, will be at a great disadvantage in the profession.

Time, practice, understanding, repetition, and concentrated effort are all required to develop listening skills. Auditory memory will increase as the listener becomes more accustomed to the sound material, more aware of patterns of levels and changes within all of the artistic elements, and more aware of how to focus on the aspects of sound that he or she was conditioned from birth to ignore. Developing refined listening skills is a long-term project; the individual's listening skills will continue throughout their professional career.

The primary skills that will be developed are the recognition of relative pitch levels and changes in levels, and relative dynamic levels and changes in levels. The ability to make certain time judgements within musical, dramatic, and clock-related contexts will also be developed throughout the method.

These skills are purposefully similar to those required of traditional musicians. The listener will need to be attentive to (*focus*) a smaller level of detail (*perspective*) of these elements than is required in traditional music listening. The listener will also need to be aware that the applications of these musical elements will be significantly different than in acoustic music. These listening skills are easily transferred, however; significant refinements to these listening skills can be accomplished remarkably quickly (with thoughtful and concentrated effort).

People often hear more than they recognize. As soon as they are able to recognize what they hear, they become better able to remember and evaluate sound (music) while listening.

An objective vocabulary and a means of evaluation have been formulated to allow precise information about sound to be extracted from the recording. The information must be an actual account of the states or values of the sound material, in order to be of use. Subjective impressions of the sound's quality (a very common way musicians and recordists attempt to communicate about sound) must be avoided, for the communication to be accurate and of value.

The physical dimensions of sound are the common experience between humans; common experiences are necessary for communication. The actual dimensions of sound need to be described, for people to communicate effectively and to exchange meaningful information about sound. Subjective impressions about a sound cannot communicate meaningful information. The perception process of hearing and the personal experiences through which humans conceptualize sound will significantly distort impressions about a sound between individuals.

To make communication meaningful, a common experience must be used as a reference for exchanging information. The only common sound experience human beings have is the sound as it exists in the air (its acoustic state), before it is separately processed by each individual. All other processes have been interpreted by the individual and are therefore distorted.

The evaluation of sound and music in this book will be a description of the actual physical dimensions and characteristics of sound, as they are processed by one's perception (as the perceived parameters of sound), and perhaps in context (as the artistic elements of sound). The evaluation process will consider the distortions of the sound caused by human perception, and account for them within any observation or description of the sound.

The contexts of a sound to be evaluated may be any application of everyday living or music recordings, or any audio application (from a workbench evaluation of the characteristics of a piece of audio equipment to a motion picture sound track).

A process of evaluating sound quality has been devised, to provide a way to communicate about sound as it actually exists. It avoids any personal impressions about the sound quality, and addresses only the physical dimensions of sound as they are perceived and as they appear in the music.

The method for evaluating sound will allow individuals to talk "about" sound in meaningful ways. It will allow them to exchange precise and conclusive information about the sound qualities, once they are able to recognize those qualities (by acquiring the above skills).

A vocabulary through which they can communicate "about" sound does not currently exist. A method of sound evaluation will be presented, along with a small vocabulary (that has the potential of generating a much larger vocabulary) "on" sound. The vocabulary must be the same for both the technical and the artistic people in the industry, to allow them to communicate effectively. Meaningless descriptions of sounds, which appear as words like "harsh," "crisp," "dirty," or "sweet," will no longer need to be used; their continued use is strongly discouraged in all contexts. They will be replaced by a description of the activity and characteristics of the sound's component parts.

The method for evaluating the dimensions of sound in music and audio recordings is equally valid for evaluating sound quality in both technical and creative contexts. It may be utilized in both critical listening and analytical listening applications.

The method may be used to communicate information about any aspect of sound. It describes the physical states of the sound's components, through a careful and thoughtful evaluation process that considers the perceived characteristics of the sound, the conditions under which the sound exists, and the conditions under which the sound is perceived. The method will represent a change of thought and listening processes for many people; readers are encouraged to continually reevaluate their own performance of sound evaluation and the ways they talk "about" sound and music recordings.

The audio recording process has provided the creative artist with increased control over the perceived parameters of sound (through direct control of the physical dimensions of sound). This increased control gives the audio recording medium resources for creative expression that are not possible acoustically or in the reality of live, acoustic performance.

New creative ideas and new additions to the musical language have emerged as a result of the audio recording process. These artistic elements are used to shape, enhance, or create music.

The Art of Recording lies in using the creative resources of music (audio) recording for artistic expression.

I

The Artistic Aspects of the Audio Recording Medium

1

The Elements of Sound in Audio Recording

Audio recording is the recording of sound. It is capturing the physical dimensions of sound and then reproducing those dimensions either immediately or from a storage medium (magnetic, solid, electronic, digital), thereby returning those dimensions to their physical, acoustic state. The process moves from physical sound, through the recording/reproduction chain, and back to physical sound.

The "art" in recording is the artistically sensitive application of the recording process to shape or create sound as, or in support of, an artistic, creative message. To be in control of crafting the artistic product, one must be in control of the recording process, the ways in which the recording process modifies sound, and communicating well-defined creative ideas.

These areas of control of the artistic process all closely involve a human interaction with sound. Inconsistencies between the various states of sound are present throughout the audio recording process. Many of these inconsistencies are the result of the human factor—the ways in which humans perceive sound and conceptualize its meanings. In order for the artist to be in control of his or her material, the artist must understand the substance of the material: sound, in all its inconsistencies.

THE STATES OF SOUND

In audio recording, sound is encountered in three different states. Each of these three states directly influences the recording process, as well as creating (or capturing) a piece of art. These three states are:

1. Sound as it exists physically (having physical dimensions).
2. Sound as it exists in human perception (psychoacoustic conception)—that is, sound being perceived by humans after being transformed by the

ear and interpreted by the mind (the perceived parameters of sound being human perceptions of the physical dimensions).

3. Sound as it exists as an aural representation of an abstract or a tangible concept, or a physical object or activity. (This is how the mind finds meaning from its attention to the perceived parameters of sound.) Sounds as meaningful events, capable of communication, provide a medium for artistic expression; sounds thereby communicate and have meaning.

The audio recording process ends with sound existing in its physical state, in air. Often, the audio recording process will begin with sound in this physical state.

Humans are directly involved in the audio recording process–they evaluate the audio signal (the recordist and all others involved in the industry), and the end listener is the reason for making the recording. Humans translate the physical dimensions of sound into the perceived parameters of sound through the listening process (aural perception).

This translation process involves the hearing mechanism functioning on the physical dimensions of sound and neural signals transmitting to the brain. This translation process is nonlinear and distorts the information; the hearing mechanism does not directly transfer acoustic energy into nerve impulses.

Certain aspects of the distortion caused by the translation process are, in general, consistent between listeners and between hearings; they are related to the physical workings of the inner ear or to the transfer of the perceived sounds to the mind/brain. Other aspects are not consistent between listeners and between hearings; they relate to the listener's particular hearing characteristics or intelligence.

The final function occurs at the brain. At a certain area of the cortex, the information is processed, identified, consciously perceived, and stored in short-term memory; the neural signals are transferred to other centers of the brain for long-term memory. At this point, the knowledge, experience, attentiveness, and intelligence of the listener become factors in perceiving of sound's artistic elements (or the meanings or message of the sound). The individual is not always sensitive or attentive to the material or to the listening activity, and is not always able to match the sound to previous experiences or known circumstances.

The physical dimensions (1.) are interpreted as perceived parameters of the sound (2.). The perceived parameters of sound (2.) provide a resource of elements that allow for communicating and understanding the meaning of sound (and artistic expression) (3.).

The audio recording process communicates ideas. The audio might be communicating music, dialog, motion picture action sounds, or whale songs, but it is producing sounds that have some type of meaning to the listener. The perceived sound provides a medium of variables that, when presented in certain orders or patterns that are recognizable, have meaning. Sound, as perceived and understood by the human mind, becomes the resource for creative and artistic expression. The perceived parameters of sound are utilized as the artistic elements of sound, to create and ensure the communication of meaningful (musical) messages.

The individual states of sound as physical dimensions and as perceived parameters will be discussed individually, immediately following. The interaction of the perceived parameters of sound will follow the discussion of the individual parameters. These discussions provide critical information for understanding the breadth of the "artistic elements of sound" in audio recording, presented in the next chapter.

PHYSICAL DIMENSIONS OF SOUND

Five physical dimensions of sound are central to the audio recording process. These physical dimensions of concern are: the characteristics of the sound waveform as (1) *frequency* and (2) *amplitude* displacements, occurring within the continuum of (3) *time*; the fusion of the many frequency and amplitude anomalies of the single sound to create a global, complex waveform as (4) *timbre*; and the interaction of the sound source (timbre) and the environment in which it exists, which creates alterations to the waveform according to the geometry of the acoustic (5) *space*.

Frequency is the number of similar, cyclical displacements in the medium (air), per time unit (measured in cycles of the waveform per second, or hertz). Each similar displacement creates a single cycle of the waveform. Amplitude is the amount of displacement of the medium at any moment, within each cycle of the waveform (measured as the magnitude of displacement in relation to a reference level, or decibels) (Figure 1-1).

Timbre is a composite of a multitude of frequency functions and amplitude displacements—the global result of all the amplitude and frequency components that create the individual sound. It is the overall quality of a sound. Its primary component parts are the dynamic envelope, spectrum, and spectral envelope.

The *dynamic envelope* of a sound is the contour of the changes in the overall dynamic level of the sound throughout its existence. Dynamic envelopes of individual acoustic instruments and voices vary greatly in content and contours. The dynamic envelope is often thought of as being divided into a number of component parts. These component parts may or

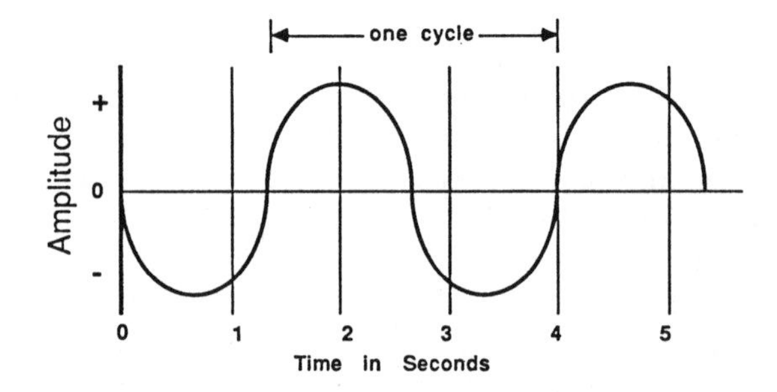

FIGURE 1-1. Dimensions of the Waveform.

may not be present in any individual sound. The widely accepted components of the dynamic envelope are: attack (time), initial decay (time), initial sustain level, secondary decay (time), primary sustain level, and final decay (release time)(Fig.1-2).

Dynamic envelope shapes, other than those created by the previous outline, are common. Many musical instruments have more or fewer parts to their characteristic dynamic envelope. Further, vocalists and performers of many instruments have great control over the sustaining portions of the envelope. Musical sounds that do not have some variation of level during the sustain portion of the envelope are rare; the organ is one such exception.

The *spectrum* of a sound is the composite of all the frequency components of the sound. It is comprised of the fundamental frequency, harmonics, and overtones.

The periodic vibration of the waveform produces the sensation of a dominant frequency. The period or cycle of the waveform will determine

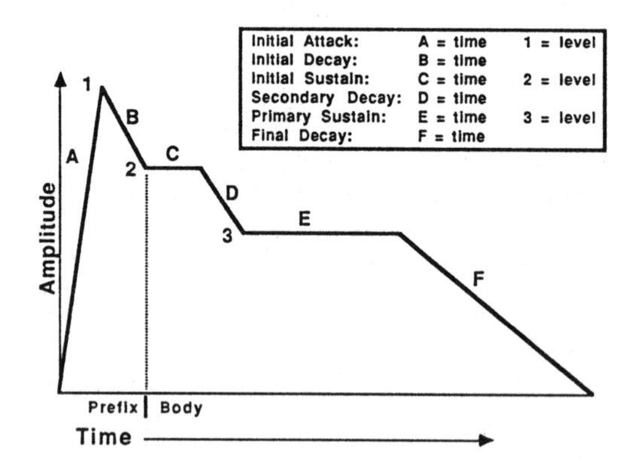

FIGURE 1-2. Dynamic Envelope.

its *fundamental frequency*; it is also that frequency at which the sounding body resonates along its entire length. The fundamental frequency is often the most prominent frequency in the spectrum. Therefore, the fundamental frequency will often have the greatest amplitude of any component of the spectrum.

In all sounds except the pure sine wave, frequencies other than the fundamental are present in the spectrum. These frequencies are primarily higher than the fundamental frequency, and may or may not be in a whole-number relationship to the fundamental. Frequency components of the spectrum that are in a whole-number relationship to the fundamental are *harmonics*; these frequencies reinforce the prominence of the fundamental frequency (and the pitched quality of the sound). Those components of the spectrum that are not proportionally related to the fundamental are *overtones*. Traditional musical acoustics studies define overtones as being proportional to the fundamental, but with a different sequence than harmonics (first overtone=second harmonic, etc.); the traditional definition has been replaced by this differentiation between *partials* that are proportional to the fundamental (harmonics) and those that are not (overtones). All the individual components of the spectrum are partials. *Partials* (overtones and harmonics) can exist below the fundamental frequency as well as above; they are referred to as sub-harmonics and subtones.

For each individual instrument or voice, certain ranges of frequencies within the spectrum will be emphasized consistently, no matter what the fundamental frequency is. Instruments and voices will have resonances that will strengthen those spectral components that fall within these definable frequency ranges. These areas, or resonance peaks, are called *formants*, or formant regions. Formants do not shift with the sound source's spectrum; they are spectral components, often noise transients (a number of frequencies, closely spaced in a "band" or group, that change in amplitude over time), that remain relatively constant throughout the sound source's range. They modify the amplitudes of the spectral content of the sound source, or appear as spectral components in themselves; they are often associated with resonances of the particular mechanism that produced the source sound.

A sound's spectrum is comprised primarily of harmonics and overtones that create a characteristic pattern that is recognizable as being characteristic of a particular instrument or voice. This pattern of spectrum will transpose (change level) with the fundamental, mostly unchanged, to form a similar timbre at a different pitch level. Formants establish frequency areas that will be emphasized or de-emphasized for a particular instrument or voice. These areas will not change with varied fundamental frequency, as they are fixed characteristics of the device that created the sound. They may also take the form of spectral information, which is present in all sounds produced by the instrument or voice.

The frequencies that comprise the spectrum (fundamental frequency, harmonics, overtones, sub-harmonics, and subtones) all have different amplitudes and different dynamic envelopes. The *spectral envelope* is the composite of each individual dynamic level and dynamic envelope of all the components of the spectrum.

The component parts of timbre (dynamic envelope, spectrum, and spectral envelope) display strikingly different characteristics during different parts of the duration of the sound. The duration of a sound is commonly divided into two time units: the *prefix* or *onset,* and the *body*. The initial portion of the sound is the prefix or onset; it is markedly different from the remainder of the sound, the body. The time length of the prefix is usually determined by the manner in which a sound is initiated, and is often the same time unit as the initial attack. The actual time increment of the prefix may be anywhere from a few microseconds to 20 to 30 milliseconds. The prefix is defined as the initial portion of the sound that has markedly different characteristics of dynamic envelope, spectrum, and spectral envelope than the remainder of the sound.

The interaction of the sound source (timbre) and the environment in which it is produced will create alterations to sound. These changes to the sound source's sound quality are created by the acoustic space. These alterations are directly related to the characteristics of the acoustic space in which the sound is produced, as well as the location of the sound source within the environment.

Space-related sound measurements must be performed at a specific, physical location. The measurements are calculated from the point in space where a receptor (perhaps a microphone, or a listener) has been placed to capture the composite sound (the sound source within the acoustic space). The location of the receptor becomes a reference in measuring the acoustic properties of space.

The aspects of space that influence sound in audio recording are (1) the *distance* of the sound source to the receptor, (2) the *geometry of the host environment* in which the sound source is sounding, (3) the *angular trajectory* of the sound source to the receptor, and(4) the *location* of the sound source *within the host environment.*

The environment in which the sound source is sounding is often referred to as the *host environment.* Within the host environment, sound will travel on a direct path to the receptor (as *direct sound*) and sound will bounce off reflective surfaces before arriving at the receptor (as *reflected sound*).

Reverberant sound is a composite of many reflections of the sound arriving at the receptor in close succession. The many reflections that comprise the reverberant sound are spaced so closely that the individual reflections cannot be perceived; the many reflections are therefore considered a single entity. As time progresses, these closely spaced reflections become even more closely spaced and of diminishing amplitude, until they are no longer of consequence. *Reverberation time* is the length of time required for the reflections to reach an amplitude level of 60 dB lower than that of the original sound source (Fig. 1-3).

Early reflections are those reflections that arrive at the receptor within 50 milliseconds of the direct sound. As a collection, the reflections that arrive at the receptor within the first 50 milliseconds after the direct sound arrives comprise the *early sound field.*

The distance of the sound source from the receptor will alter the composition of the sound at the receptor. The sound at the receptor will

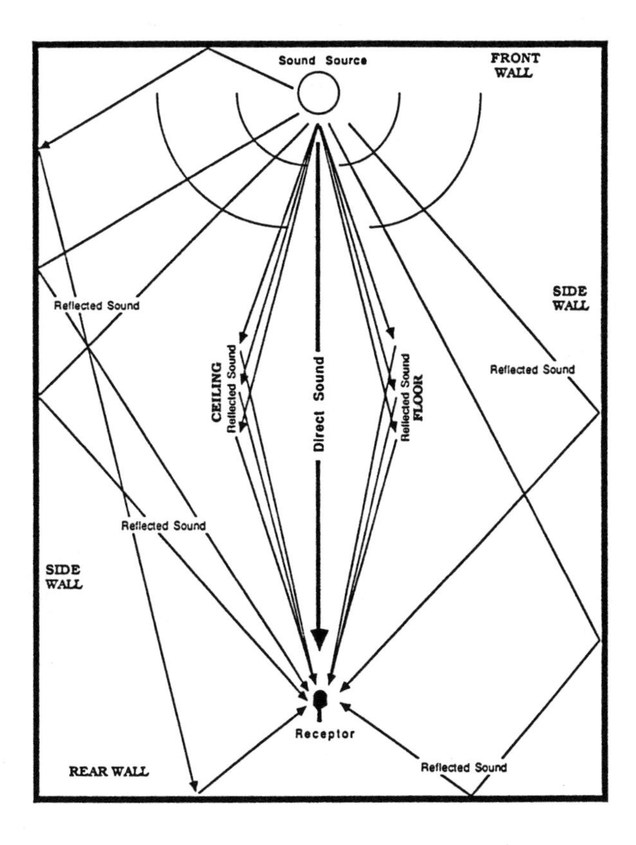

FIGURE 1-3. Paths of Sound Within an Enclosed Space.

be a composite of the direct sound and the reflected sounds (reverberation and early sound field). The composite sound at the receptor is affected by the distance of the sound source from the receptor in two ways:

1. The ratio of the amplitude of the direct sound to the amplitude of the reflected sound becomes more equally distributed, as the distance of the sound source to the receptor increases.
2. Low amplitude portions of the sound's spectrum (usually high frequencies) are lost with increasing distance of the sound source to the receptor.

The characteristic changes to the composite sound caused by the geometry of the host environment, the angular trajectory of the sound source to the receptor, and the location of the sound source within the host environment are related to the changes caused by distance, with the above two variables to the composite sound remaining as influences.

These three dimensions of the relationship of the sound source to its acoustic space may alter the composite sound in the following, additional ways:

1. Timbre differences between the direct and reflected sounds (especially important for characteristics of the host environment);
2. Time differences between the arrivals of the direct sound and the initial reflections (especially important to the angle of the sound source from the receptor and the relationship of the sound source to the reflective surfaces of the host environment);
3. Spacing in time of the early reflections (especially important to the angle of the sound source from the receptor and the relationship of the sound source to the reflective surfaces of the host environment); and
4. Amplitude differences between direct and reflected sounds (especially important to the angle of the sound source from the receptor and for characteristics of the geometry of the host environment).

The geometry of the host environment greatly influences the content of the composite sound. The dimensions and volume of the space, the trajectories and materials of construction of the boundaries (walls, floors, ceilings), and the composition and content (such as windows and large objects within the space) of the host environment all alter the composite sound. Host environments cover the gamut of all the physical spaces and open areas that create reality (from a small room to a large concert hall; from the corridor of a city street to an open field; etc.).

In audio production, the spatial properties of host environments can be generated artificially; the commonly used possibility exists to alter the composite sound in ways that simulate environmental characteristics that cannot occur in physical reality.

As stated earlier, the characteristics of the composite sound are influenced by the angular trajectory of the sound source to the receptor. The sound source may be at any angle from the receptor and be detected. It may be present at any location in relation to the receptor. The location is calculated with reference to juxtaposed 360-degree spheres that encompass the receptor.

The calculation of the angle of the sound source to the receptor may utilize the spheres that are parallel to the floor of the space as a reference, thereby placing the sound source and the receptor on the same *horizontal plane*. The calculation of the angle of the sound source to the receptor may utilize the spheres that are at a 90-degree angle to the floor of the space, thereby placing the sound source and the receptor on the same *vertical plane*. The precise location of the sound source within a three-dimensional space is determined by defining the sphere that will serve as a reference and then defining the angle of the sound source to the receptor within the reference sphere (Fig. 1-4).

The location of the sound source within the host environment will cause unique sequences of reflected sound (when the source is producing sound within the environment). These unique sequences will be comprised of patterns of reflections as they are spaced over time (a "rhythm of reflec-

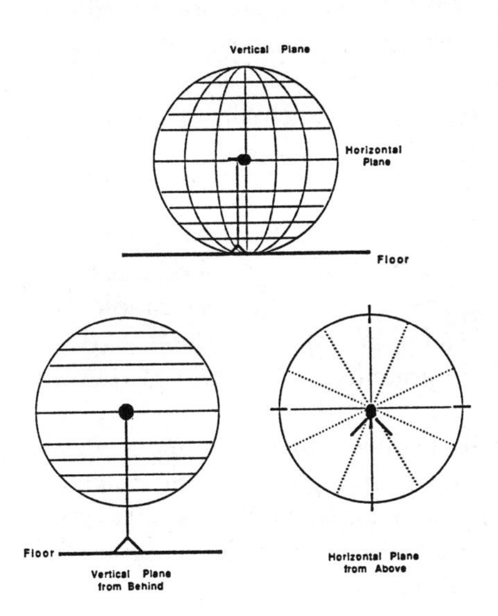

FIGURE 1-4. Defining the Angular Trajectory of the Sound Source from the Receptor.

tions"), and patterns of the amplitudes of the reflected sounds within the reverberant sound and in relation to the direct sound. By generating unique early time field and reverberant sound, the location of the sound source within the host environment may cause significant alterations to the composite sound at the receptor. The amount of influence that the location of the sound source, within the host environment, will have on the content of the composite sound will be directly related to the proximity of the sound source to the walls, ceiling, floor, openings (such as windows and doors), and large objects that reflect sound within the host environment.

PERCEIVED PARAMETERS OF SOUND

The five physical dimensions of sound translate into respective perceived parameters of sound. The state of sound as it exists in human perception is quite different from the physical state of sound. People's perception of sound is a result of the physical dimensions being transformed by the ear and interpreted by the mind. The perceived parameters of sound are human perceptions of the physical dimensions of sound.

This translation process from the physical dimensions to the perceived parameters is nonlinear. The hearing mechanism does not directly transfer acoustic energy into equivalent nerve impulses. The human ear is not equally sensitive in all frequency ranges, nor is it equally sensitive to sound at all amplitude levels. This nonlinearity of acoustic energy transfer to neural impulses causes the physical dimensions of sound to be in a different state in people's perception than actually exists in air.

Physical Dimension	*Perceived Parameter*
Frequency	Pitch
Amplitude	Loudness
Time	Duration
Timbre (physical components)	Timbre (perceived overall quality)
Space (physical components)	Space (perceived quality)

Pitch is the perceived conception of frequency. The frequency area most widely accepted as encompassing the hearing range of the normal human spans the boundaries of 20 Hz to 20,000 Hz (20 kHz).

Most humans cannot identify specific pitch levels. Some people have been blessed with an innate ability of being able to recognize specific pitch levels (in relation to specific tuning systems), and some people have developed this ability. These people are said to have "absolute pitch," or "perfect pitch." The ability to quickly and accurately recognize precise pitch levels is not common even among well-trained musicians.

It is within human ability, however, to determine the relative placement of a pitch within the hearing range. A *register* is a specific portion of the *range*. It is entirely possible to determine, within certain limits of accuracy, the relative register of a perceived pitch level. This skill can be developed and accuracy improved.

Humans are able to consistently estimate the approximate level of a pitch, associating pitch level with register. With practice, this consistency can be accurate to within a minor third (three semi-tones). This skill in the "estimation of pitch level" will be important in evaluating sound.

Humans perceive pitch most accurately as the relationship between two or more soundings of the same, or associated, sound sources, not as identifiable, discrete increments. They do not listen to pitch material to define the letter-names (increments) of pitches. Instead, humans calculate the distance (or interval) between pitches by gauging the distance between the perceived levels of the two (or more) pitches.

The interval between pitches becomes the basis for all judgements that define and relate the sounds. Thus, melody is the perception of successively sounded pitches (creating linear intervals), and chords are the perceptions of simultaneously sounded pitches (creating simultaneous, or harmonic intervals). Pitch is often perceived in relation to a reference level (one predominating pitch that acts as the key or pitch-center of a piece of music) or to a system of organization to which pitches can be related (a tonal system, such as major or minor).

The ability to recognize the interval between two pitches is not consistent throughout the hearing range. Most listeners have the ability to accurately judge the size of the semi-tone (or minor second, the smallest musical interval of the equal tempered system) within the range of 60 Hz and 4 kHz. As pitch material moves below 60 Hz, the normal listener will have increased difficulty in accurately judging interval size. As pitch material moves above 4 kHz, the normal listener will have increased difficulty in accurately judging interval size.

The smallest interval humans can accurately perceive varies with the register and placement of the two pitches creating the interval. The size of the minimum audible interval varies from about one-twelfth of a semi-tone between 1 to 4 kHz to about one-half of a semi-tone (a quarter-tone) at approximately 65 Hz. These figures are dependent upon optimum duration and loudness levels of the pitches; sudden changes of pitch level are up to 30 times easier to detect than gradual changes. It is possible for humans to distinguish up to 1500 individual pitch levels by spacing out the appropriate minimum-audible intervals throughout the hearing range.

With all factors being equal, the perception of harmonic intervals (simultaneously sounding pitches) is more accurate than the perception of melodic intervals (successively sounded pitches). Up to approximately 500 Hz, melodic and harmonic intervals are perceived equally well. Above 1 kHz, humans are able to judge harmonic intervals with greater accuracy than melodic intervals. Above 3500 Hz, this difference becomes pronounced.

Loudness is the perception of the overall excursion of the waveform (amplitude) at any moment in time. Amplitude can be physically measured as sound pressure level. In perception, loudness level cannot be accurately perceived in discrete levels.

Loudness is referred to in relative values, not as having separate and distinct levels of value. Traditionally, loudness levels have been described by analogy ("louder than," "softer than," etc.) or by relative values ("soft," "medium loud," "very soft," "extremely loud," etc.). Humans compare loudness levels, conceiving them as being "louder than" or "softer than" the previous, succeeding, or remembered loudness level(s).

A great difference exists between loudness as perceived by humans and the physical amplitude of the sound wave. This difference can be quite large at certain frequencies. In order for a sound of 20 Hz to be audible, a sound pressure level of 75 decibels must be present. At 1 kHz, the human ear will perceive the sound with a minute amount of sound pressure level, and at 10 kHz a sound pressure level of approximately 18 dB is required for audibility.

The nonlinear frequency response of the ear and the fatigue of the hearing mechanism over time play an important role in further inaccuracies of human perception of loudness. With sounds of long durations and steady loudness level, loudness will be perceived as increasing with the progression of the sound, until approximately 0.2 seconds of duration. At this time, the gradual fatigue of the ear (and possibly shifts of attention of the listener) will cause perceived loudness to diminish.

As the loudness level of the sound is increased, the ear requires increasingly more time between soundings, before it can accurately judge the loudness level of a succeeding sound. People are unable to accurately judge the individual loudness levels of a sequence of high-intensity sounds as accurately as they can judge the individual loudness levels of mid- to low-intensity sounds. The inner ear itself requires time to reestablish a state of normalcy, from which it can accurately track the next sound level.

As a sound of long duration is being sustained, its perceived loudness level will gradually diminish. This is especially true for sounds that have high sound pressure levels. The ear gradually becomes desensitized to the loudness level. The physical masking (covering) of softer sounds and an inability to accurately judge changes in loudness levels will result from the fatigue. When the listener is hearing under listening fatigue, slight changes of loudness may be judged as being large. Listening fatigue may desensitize the ear's ability to detect frequencies within the frequency band (frequency area), where the high sound pressure level was present.

Humans perceive *time* as *duration*. Durations of sounds are not perceived individually. Humans cannot accurately judge time increments, without a reference time unit. Regular reference time units are found in musical contexts, but rarely in other aspects of human existence. Even the human heartbeat is not consistent enough to act as a reliable reference. The underlying metric pulse of a piece of music allows for an accuracy in duration perception. This accuracy cannot be achieved in any other context of human experience.

The listener remembers the relative duration values of successive sounds, in a process similar to that of perceiving melodic pitch intervals. These successive durations create musical rhythm. The listener calculates the length of time between when a sound starts and when it ends, in relation to what precedes it, what succeeds it, what occurs simultaneously with it, and what is known (what has been remembered). Instead of calculating a span of pitch, the listener is attempting to calculate a span of time, as a durational value.

A *metric grid*, or an underlying pulse, is quickly perceived by the listener, as a piece of music unfolds. This metric grid provides a reference pulse, to which all durations are related. The listener is thereby able to make rhythmic judgements in a precise and consistent manner. The equal divisions of the grid allow the listener to compare all durations and to calculate the pulse-related values of the perceived sounds. Durations are calculated as being in proportion to the underlying pulse: at the pulse, half a pulse, quarter of the pulse, double the pulse, and so on.

In the absence of the metric grid, durational values cannot be accurately perceived as proportional ratios. Humans will not be able to perceive slight differences in duration, when a metric grid is not established.

The listener is able to establish a metric grid only within certain limits. Humans will be able to accurately utilize the metric grid, with accuracy, between 30 to 260 pulses per minute. Beyond these boundaries, the pulse is not perceived as the primary underlying division of the grid. The human mind will replace the pulse with a duration of either one-half or twice the value, or the listener will become confused and unable to make sense of the rhythmic activity.

The metric grid is the dominant factor in perceiving tempo, as well as musical rhythm. In most instances, the metric grid itself represents the steady pulsation of the tempo of a piece of music.

The listener's *perception of time* plays a peripheral role in perceiving rhythm. Time perception is significant in perceiving the global qualities of a piece of music, and in estimating durations, when a metric grid is not present in the music. The global qualities of aesthetic, communicative, and extra-musical ideas within a piece of music depend largely on the living experience of music, on the passage of musical materials across the listener's time perception of his or her existence.

Time perception is distinctly different from duration perception. The human mind bases time judgements on the amount and nature of the musical material heard within the "length of the present." The length of time humans perceive to be "the present" is normally two to three seconds, but might be extended to as much as five seconds.

The "present" is the window of consciousness, through which people perceive the world and listen to sound. They are at once experiencing the moment of their existence, evaluating the immediate past of what has just happened, and anticipating the future (projecting what will follow the present moment, given their experiences of the recently passed moments, and their knowledge of previous, similar events).

Human time judgements are imprecise. The speed at which events take place and the amount of information that takes place within the "present" greatly influences time judgements. The amount of time perceived to have passed will change, to conform to the number of events experienced within the present. The listener will estimate the amount of time passed in relation to the number of experiences during the present, and make time judgements accordingly.

Time judgements are greatly influenced by the individual listener's attentiveness and interest in what is being heard. If the material stimulates thought within the listener, the event will seem shorter; if the listener finds the listening activity desirable in some way, the experience will seem to occupy less time than would an undesirable experience of the same (or even shorter) length. Expectations caused by, boredom with, interest in, contemplation of, and even pleasure created by music alter the listener's sense of elapsed time.

The "time length" of a piece of music (or any time-based art form, such as a motion picture) is separate and distinct from clock time. A lifetime can pass in a moment, by experiencing a work of art. A brief moment of sound might elevate the listener to extend the experience to an infinity of existence.

The overall quality of a sound, its timbre, is the perception of a composite of all the physical aspects that comprise a sound. Timbre is the global form, or the overall character of a sound, that one can recognize as being unique.

The overall form (timbre) is perceived by the states and interactions of its component parts. The physical dimensions of sound, as discussed earlier, are perceived as dynamic envelope, spectral content, and spectral envelope (perceived values, not physical values). The perceived dimensions are interpreted, and they shape an overall quality, or conception of the sound.

Humans remember timbres as conceptions, as entire objects having an overall quality (that is, comprised of unique characteristics), and sometimes as having meaning in itself (as a timbre can bring with it associations in the mind of the listener). People remember the timbres of hundreds of human voices and the timbres of a multitude of sounds from their living experiences. They remember the timbres of many musical instruments, and their different timbres as they are being performed in many different ways.

The global quality, or *Gestalt*, of the perceived sound (that is, timbre) allows us to remember and recognize specific timbres as unique and identifiable objects.

Humans have the ability to recognize and remember a large number of timbres. Further, listeners have the ability to scan timbres and relate unknown sounds to sounds stored in the listener's long-term memory. The listener is then able to make accurate and meaningful comparisons of the states and values of the component parts of the timbres. These skills will serve as meaningful points of departure for the method of evaluating timbre in Part 3.

Perceiving timbre requires sufficient time for the mind to process the global characteristics of the sound. The time required to perceive the component parts of timbre vary significantly with the complexity of the sound and the listener's previous exposure to the timbre. For rather simple sounds, the time required for accurate perception is approximately 60 milliseconds. As the complexity of the sound is increased, the time needed to perceive the sound's component parts will also increase. All sounds with a duration of less than 50 milliseconds will be perceived as being noise-like, as the identification of a specific timbre is not possible at that short of a duration; exceptions occur only when the listener is very well acquainted with the sound, and the timbre is recognized from this small bit of information.

The components of the timbre's spectrum (the partials) fuse to create the impression of a single sound. Although many frequencies are present, the tendency of our perception is to combine the many sounds into one overall texture. Fusion occurs in such a way that it may be possible to discern the lower partials of a complex sound, but the higher partials (generally of a much lower loudness level) are much more difficult to detect. Fusion is so strong a part of people's experience that they fuse partials that are harmonically proportional to the fundamental frequency, as well as overtones that are distantly related to the fundamental.

Fusion can also occur between two separate timbres (two individual sound sources), if the proper conditions are present. Timbres that are attacked simultaneously, or that are of a close harmonic relationship to each other, are the most likely to fuse into a perceived composite sound. The more complex the individual sound, the more likely that fusion will *not* occur. Furthermore, if the listener recognizes one of the timbres, fusion will be less likely to occur; the listener may quickly segregate both sounds. Again related to recognizing a timbre, synthesized sounds are more likely to fuse than are sounds of an acoustic origin.

The perception of the *spatial characteristics* of sound is the conception of the physical location of a sound source in an environment, and the modifications that the environment itself places on the sound source.

The perception of space in audio recording (reproduction) is not the same as the perception of space of an acoustic source in a physical environment. In an acoustic space, listeners perceive the location of sound in relation to the three-dimensional space around them. Sound is perceived at any angle from the listener, and sound is perceived at a distance from the listener; both of these perceptions involve evaluating the characteristics of the sound source's host environment.

In audio recording, illusions of space are created. Sound sources are assigned spatial information (through the recording process and/or through signal processing). The spatial information is intended to simulate particular known, physical environments or activities, or is intended to provide spatial cues that have no relation to our reality. Theoretically, all of the interactions of the sound with its host environment are captured with, or can be simulated and applied to, the sound source. Upon playback through two loudspeakers, the spatial cues are to be accurately reproduced.

The sound sources (with spatial characteristics) will be heard by the listener through two loudspeakers. Further complicating matters, the loudspeakers themselves are placed in and interact with a physical environment (an environment that is quite unrelated to the spaces on the recording). The spatial characteristics applied to the sound source are perceived by the listener after they have been distorted by the characteristics of the loudspeakers, distorted by the placement of the loudspeakers within the playback environment, and distorted by the playback environment itself. The listener perceives the reproduced spatial characteristics of the sound source within the three-dimensional space of the listening environment (headphone monitoring is not a solution, as will be discussed in Chapter 5).

To accurately perceive the spatial information of an audio recording, the listening environment must be acoustically neutral, and the listener must be carefully positioned both within the environment and in relation to the loudspeakers. The listening environment (including the loudspeakers) must not place additional spatial cues onto the reproduced sound.

Humans perceive spatial relationships as the location of the sound source being at an angle to the listener (above, below, behind, to the left, to the right, in front, etc.), as the location of the sound source being at distance from the listener, as the location of the sound source being within an environment, and as a conceptual image of the type, size, and properties of the host environment.

These perceptions are transferred into the recording medium, to provide a realistic illusion of space, with one major exception. The angular location is severely restricted in audio reproduction, as compared to human perceptual abilities. Audio playback can accurately and consistently reproduce localization cues only on the horizontal plane, and then only slightly beyond the loudspeaker array (Fig. 1-5). The three-dimensional space of our reality is simulated (with dubious accuracy) in the two dimensions of audio recording.

Much research and development is taking place in recording and playback systems, in the attempt to extend the localization of sound sources to the vertical plane, and to behind the listener, as well as to provide a more realistic reproduction of distance and environmental cues. Significant advances are being made, and these systems hold much promise for the future. Controlling sound localization on the vertical plane and more completely simulating environmental characteristics is not, however, an available resource for most recordists, and it is certainly not available in a vast majority of the listening environments used for audio playback.

The following discussions of the perception of the spatial dimensions of reproduced sound refer to the common two-channel, listening room playback systems. These concepts of perceptions will transfer to other systems, such as quadraphonic sound, binaural recordings, holophonics, surround sound, and others, but with different boundaries of the limits of perception inherent with each particular format. The evaluation of the spatial charac-

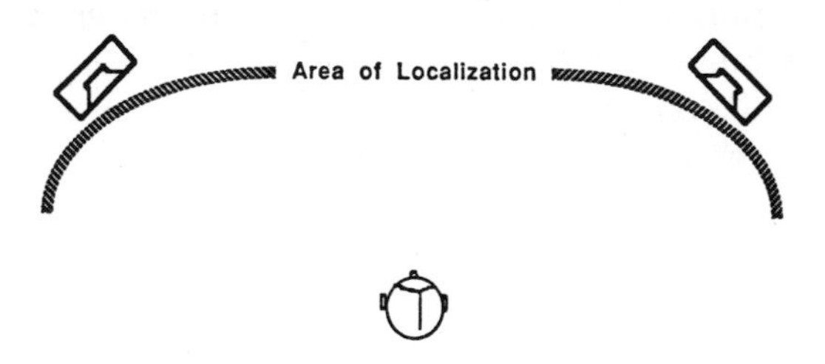

FIGURE 1-5. Area of Sound Localization in Two-Channel Audio Playback.

teristics of sound sources presented in Chapter 12 will address only the current common practice of two-channel stereo playback.

Humans use differences in the sound wave, as it appears at the two ears, for the accurate *localization of direction. Interaural time differences* (ITD) are the result of the sound arriving at each ear at a different time. A sound that is not precisely in front or in back of the listener will arrive at the ear closest to the source before it reaches the furthest ear. These time differences are sometimes referred to as phase differences. The sounds arriving at each ear are almost identical during the initial moments of the sound, except that the sound at each ear is at a different point in the waveform's cycle.

Interaural amplitude differences (IAD) work in conjunction with ITD in the localization of the direction of the sound source. Internal amplitude differences are referred to as interaural spectral differences, in some contexts. They are the result of sound pressure level differences at high frequencies present at the two ears. These differences are caused by the head of the listener, which blocks certain frequencies from the furthest ear (when the sound is not centered). This occurrence has been termed the "shadow effect." Interaural amplitude differences will at times consist solely of amplitude differences between the two ears, with the spectral content of the waveform being the same at both ears.

The sound wave is almost always different at each ear. The differences between the sound waves may be either time/phase-related, amplitude/spectrum-related, or both. These differences in the waveforms are essential for perceiving the direction of the sound source. In addition, these same cues play a major role in perceiving the characteristics of the host environment.

Up to approximately 800 Hz, humans rely on ITD for localization cues. Phase differences are utilized for localization perception, up to about 800 Hz, as amplitude appears to be the same at both ears.

Between 800 Hz and about 2 kHz, both phase and amplitude differences are present between the two ears. Both IAD and ITD are used for perceiving direction in this frequency range, with time/phase differences appearing to dominate.

In general, time/phase differences seem to dominate the perception of direction, up to about 4 kHz. Although IAD are present, ITD dominate the perception of direction between 2 kHz and 4 kHz. Humans have poor localization ability for sounds in this frequency band.

Above 4 kHz, IAD determine the perception of localization cues. Localization ability improves at 4 kHz, and is quite accurate throughout the upper registers of the human hearing range.

As seen, humans do not perceive direction accurately at all frequencies.

Below approximately 500 Hz, their perception of the angle of the sound source becomes increasingly inaccurate, to the point where sounds seem to have no apparent, focused location. An area exists around 3 kHz where localization is also poor; wave length similarities in the distance between the two ears and those of the frequencies around 3 kHz cause time interaural differences to be unstable.

Humans have a well-refined ability to localize sounds in the approximate frequency areas: 500 Hz to 2 kHz, 4 kHz–upper threshold of hearing. Within these areas, the minimum audible angle is approximately one to two degrees. Sounds that have fundamental frequencies outside of these frequency areas, but that have considerable spectral content within these bands, will also be localized quite accurately. The ability for humans to localize sounds is one of the survival mechanisms that has remained from the ancient past. This ability has been developed throughout human evolution; humans have learned to perceive the direction of even those sounds that their hearing mechanism has difficulty processing.

The perceived distance of a sound source from the listener is determined primarily by two calculations: (1) the ratio of the amount of direct sound to reverberant sound, and, the primary determinant, (2) the loss of low amplitude (usually high frequency) portions of the sound's spectrum with increasing distance (*definition of timbre*). Both of these functions rely on the listener's knowledge of the sound's timbre, for accurate perception of distance-location.

The listener must know the timbre of a sound, to recognize that the sound is missing certain low energy components of its spectrum or that the sound is missing detail. The listener will further calculate how much low energy information is missing, to determine the degree of distance.

Knowing the timbre of the sound source will assist the listener in perceiving the reiterations of the direct sound and the reverberant sound. This perception will assist in calculating the amount of distance between the sound source and the listener. Previous experiences and the listening skill level of the listener will play a major role in determining the accuracy with which distance judgements are made.

Without prior knowledge of the timbre of a sound, perception of distance location is considerably less accurate.

Related to the ratio of direct to reflected sound, as distance increases, the time difference between the cessation of the direct sound and the cessation of the reverberant energy will also increase. Through "temporal fusion," the reverberant sound is perceived as being a part of the direct sound, thus creating an overall conception of the sound in its environment (referred to as the "composite sound," above). As distance increases, temporal fusion begins to diminish and the end of the direct sound and the continuance of the reverberant energy becomes more prominent.

The perceptions of the *characteristics of the host environment* and the *placement of the sound source within the host environment* are also dependent upon the ratio of direct to reflected sound and the loss of low-level spectral components, with increasing distance. In addition, the characteristics of the host environment are perceived through:

1. The time difference between the direct sound arriving and the initial reflections arriving;
2. The spacing in time of the early reflections;
3. Amplitude differences between the direct sound and all reflected sound (the individual initial reflections and the reverberant sound); and
4. Timbre differences between the direct sound, the initial reflections, and the reverberant sound.

The time delay between the direct and the reflected sounds is directly related to:

1. The distance between the sound source and the listener;
2. The distance between the sound source and the reflective surfaces (which send the reflected sound to the listener); and
3. The distance of the reflective surfaces from the listener.

These three physical distances also create the patterns of time relationships (the "rhythms") of the early reflections.

Early reflections arrive at the listener within 50 milliseconds of the direct sound. These early reflections comprise the early sound field. The early sound field is composed of the first few reflections that reach the listener before the beginning of the diffused, reverberant sound. Many of the characteristics of a host environment are disclosed during this initial portion of the sound. The early sound field contains information that provides clues as to the size of the environment, the type and angles of the reflective surfaces, and even the construction materials used in the space.

Humans have the ability to accurately judge the size and characteristics of the host environments of sound sources merely by evaluating their sound qualities. Humans experience and remember the sound qualities of a great many natural environments (in much the same way as they remember timbres). Further, they have the ability to match those environments in their previous experiences to new environments they encounter, thereby allowing for the evaluation of the characteristics of host environments.

These listening skills of evaluating and recognizing environmental characteristics can be developed to a highly refined level. People who have studied acoustical environments have refined this ability to the point where they can perceive the dimensions of an environment, openings within an environment (doors, windows, etc.), and reflective surfaces within an environment, and can accurately determine the construction material incorporated into the various reflective surfaces of an environment.

INTERACTION OF THE PERCEIVED PARAMETERS

Perceiving any parameter of sound is always dependent upon the current states of the other parameters. Altering the state of one of the perceived parameters of sound will cause an alteration in the perceived state of at least one other parameter of sound. The parameters of sound interact, causing the perception of the state of one parameter to be altered by the state of another.

Sufficient duration is required for the ear to perceive pitch. If the duration is too short, the sound will be perceived as having indefinite pitch and as being noise-like. The time necessary for the mind to determine the pitch of a sound is dependent on the frequency of the sound. Sounds lower than 500 Hz and higher than 4 kHz require more time to establish pitch quality than sounds pitched between 500 Hz and 4 kHz. At the extremes of the hearing range, pitch quality may require as much as 60 milliseconds to become established. In the mid-frequency range, and under appropriate conditions, pitch may be perceived in as little as 3 milliseconds. The median time required to establish pitch perception appears to be 10 milliseconds.

The length of time required to establish a perception of pitch will also depend on the sound's attack characteristics and its spectral content (its timbre). Sounds with complex (but mostly harmonic) spectra and sounds that have short attack times will establish a perception of pitch sooner than other sounds.

Loudness will influence pitch perception, as humans will perceive a change of pitch with a change of loudness (dynamic) level. Below 1 kHz, a substantial increase in loudness level will cause an apparent lowering of the pitch level. The sound will appear to "go flat," although no actual change of pitch level has occurred. Similarly, a substantial increase in the dynamic (loudness) level of a pitch above 2 kHz will cause the sound to appear to "go sharp"; an impression of the raising of the pitch level is created, although the actual pitch level of the sound has remained unaltered.

Loudness level can influence perceived *time relationships*. When two sounds begin simultaneously, they will appear to have staggered entrances if one of the two sounds is significantly louder than the other. The louder sound will be perceived as having been started first.

Perceived loudness level is often distorted by the speed at which information is processed. When a large number of sounds occur in a short period of time, the listener will perceive those sounds as having a higher loudness level than sounds of the same sound pressure level that are distributed over a longer period of time. This distortion of loudness level is caused by the amount of information being processed within a specific period of time (the time period is related to the perceived "length of the present").

Masking occurs when a sound is not perceived because of the qualities of another sound. The simultaneous sounding of two or more sounds can cause a sound of lower loudness level, or a sound of more simple spectral content, to be masked or hidden from the listener's perception. The masking of sounds is a common error when people are beginning their studies in audio recording.

When two simultaneous sounds of relatively simple spectral content have close fundamental frequencies, the sounds will tend to mask each other and blend into a single, perceived sound. As the two sounds become separated in frequency, the masking will become less pronounced, until both sounds are clearly distinguishable.

Sounds of relatively simple spectral content tend to mask sounds of higher frequency. This masking becomes more pronounced as the loudness of the lower sound is increased, and is more likely to occur when a large interval separates the two pitch levels of the sounds. This masking is especially prominent if the two pitch levels are in a simple harmonic relationship (especially 2:1, 3:1, 4:1, and 6:1). A higher pitched sound can mask a lower pitched sound if the higher sound is significantly higher in loudness level and given the same conditions as above; the higher the loudness level, the broader the range of frequencies a sound can mask.

Masking can occur between successive sounds. With sounds separated in time by up to 20 to 30 milliseconds, the second sound may not be perceived if the initial sound is loud enough to fatigue the ear. In a similar way, a sound may not be perceived if it is followed by another sound of great intensity within 10 milliseconds.

Audio equipment can produce "white" or broadband noise that can mask sounds of all frequencies. An entire program might be masked by the noise of the sound system itself, should the loudness level of the noise be sufficiently higher than that of the program. This type of masking problem will first be noticed in the high frequencies, where low loudness levels exist in the upper components of the sound's spectrum.

Duration can distort the perception of loudness. Humans tend to average loudness levels over a time period of about two-tenths of a second. Sounds of shorter durations will appear to be louder than sounds (of the same intensity) of durations longer than two-tenths of a second.

Timbre influences perceived loudness. Sounds with complex spectra will be perceived as being louder than sounds that contain fewer partials or a greater number of proportionally-related partials (harmonics), with all other factors being equal. A change of timbre during the sustained duration of a sound will result in a perceived change in loudness, following this principle.

As a product of the interaction of the harmonics and closely related overtones of a sound's spectrum, a timbre can create a perception of pitch, where a fundamental frequency is not physically present. The harmonics of a sound reinforce its fundamental frequency, to enhance the perception of pitch. This phenomenon is so capable of producing the perception of the fundamental frequency that an harmonic spectrum can provide the perception of pitch when the fundamental frequency is not physically present (missing fundamental). A perception of the periodicity of the fundamental frequency is created by the spectrum of the sound, although the frequency itself may not be present.

The time between repetitions and the amplitude of reiterated sounds can influence location perception. The "precedence" or "Haas effect" results when two loudspeakers reproduce the same sound source in close succession. The effect works against the principle that, when two loudspeakers reproduce a sound simultaneously and at the same amplitude, the sound appears to be centered between the two loudspeakers.

When two loudspeakers reproduce the same sound source in close succession, normal perception would localize a sound source at the earliest sounding loudspeaker and then shift the image to center, when the second loudspeaker is sounded. The Haas effect continues the localization of the sound source at the first speaker location, while adding the second loudspeaker (to reinforce the sound intensity of the first speaker), without losing the localization of the sound source at the location of the first loudspeaker. The time difference between the sounding of the loudspeakers must be at least 3 milliseconds, to pull the sound from the center image back to the loudspeaker of origin, with 5 milliseconds being a more effective minimum. A maximum delay of no more than approximately 25 to 30 milliseconds may be utilized before the delayed signal is in danger of being perceived as an *echo* (an event that will take place at all frequencies at 50 milliseconds). If the leading channel is lowered by 10 dB, or the following channel increased by 10 dB, the sound source will again be centered.

As sounds increase in distance, more and more low-amplitude components of the spectrum are not perceived. These components of timbre are

vital for recognizing a sound source. Therefore, sounds become less distinguishable as distance increases.

The three states of sound that are utilized in audio recording are the physical dimensions of sound, the perceived parameters of sound, and sound as a resource of artistic elements. The physical dimensions of sound are interpreted as the perceived parameters of the sound. The perceived parameters of sound are transmitted to the brain and provide a resource of elements that allow for the communication and understanding of the meaning of sound (and artistic expression).

The final function occurs at the mind/brain. The brain perceives a work of art as information processed, identified, and understood, and as an experience. The perceived sound is a medium for variables that, when presented in certain orders or patterns that are recognizable, provides the sound event with meaning.

Sound as it is perceived and understood by the human mind becomes the resource for creative and artistic expression in sound. The perceived parameters of sound are utilized as the "artistic elements of sound," to create and ensure the communication of meaningful (musical) messages. The "artistic elements of sound" in audio recording are presented in the next chapter.

2

The Artistic Elements of Recording Production

The audio recording process has provided the creative artist with a very precise control over the perceived parameters of sound (through a direct control of the physical dimensions of sound). This control of sound is well beyond that which was available to composers and performers before the advances of recording technology. The potential of controlling sound in new ways has led to new artistic elements in music, as well as to a redefinition of the musician and of music. While this discussion focuses on musical applications, it must be remembered that all aspects of these artistic elements of music also function as artistic elements in other areas of audio.

A new creative artist has evolved. This person uses the potentials of recording technology as sound resources for the creation (or recreation) of an artistic product. This person may be a performer or composer in the traditional sense, or may be one of the new musicians–a producer, sound engineer, or any of the host of other, related job titles. Throughout this book, these people are referred to as "recordists."

Through its detailed control of sound, the audio recording medium has resources for creative expression that are not possible acoustically. Sounds are created, altered, and combined in ways that are beyond the reality of live, acoustic performance. New, creative ideas and new additions to the musical language have emerged as a result of the audio recording process.

Creative ideas are defined by the "artistic elements." The artistic elements are the areas that comprise or characterize creative ideas (or entire works of art, pieces of music). Studying the artistic elements allows one to understand the individual musical ideas and the larger musical event, and recognize how those ideas and sound events contribute to the entire piece of music. Discussion will emphasize the artistic elements that are unique to recorded music, especially music created through the use of the recording medium.

As stated in the previous chapter, the artistic elements of sound are the mind/brain's interpretation of the perceived parameters of sound. Sound as it is perceived and understood by the human mind becomes the resource for creative and artistic expression. The perceived parameters of sound are utilized as the "artistic elements of sound," to create and ensure the communication of meaningful (musical) messages.

The Art of Recording lies within the utilization of the perceived parameters of sound as a resource for artistic expression. The materials that allow for artistic expression are understood through a study of their component parts—the artistic elements of sound.

By applying the perceived parameters of sound to the creation of a recording, the recorded material is comprised of sound elements that are interpreted by the mind/brain, and thus communicate artistic ideas. The artistic elements are directly related to specific perceived parameters of sound, as the perceived parameters of sound were directly related specific physical dimensions of sound.

As will be remembered, sound in audio recording is in three states: physical dimensions, perceived parameters, and artistic elements.

The artistic elements are the recordist's resources for artistic expression. The perceived parameters translate into the artistic elements:

1. Pitch becomes pitch levels and relationships.
2. Loudness becomes dynamic levels and relationships.
3. Duration becomes rhythmic patterns and rates of activities.
4. Timbre becomes sound sources and sound quality.
5. Space becomes spatial properties.

The audio production process provides the resources for considerable variation, as well as the very refined control of *all* of the artistic elements of sound. This allows all the artistic elements of sound to be accurately and precisely controlled through many states of variation, in ways that were possible with *only* pitch on traditional musical instruments (Table 2-1).

PITCH LEVELS AND RELATIONSHIPS

Relationships of pitch levels contain most of the pertinent information in a piece of music. The artistic message of most of the music of Western heritage is communicated (to a large extent) by pitch relationships. The listener has been trained, by the music heard throughout his or her life, to focus on this element in order to obtain the most significant musical information. The other artistic elements often support pitch patterns and relationships.

Pitch is the most precisely controlled artistic element in traditional music. The use of pitch relationships and pitch levels in music is more sophisticated

TABLE 2-1. The States of Sound in Audio Recording

Physical Dimension (Acoustic State)	Perceived Parameters (Psychoacoustic Conception)	Artistic Elements (Resources for Artistic Expression)
Frequency	Pitch	Pitch Levels and Relationships melodic lines, chords, register, range, tonal organization, textural density, pitch areas, vibrato
Amplitude	Loudness	Dynamic Levels and Relationships dynamic contour, emphasis accent, tremolo, balance (dynamic relationships of sound sources)
Time	Duration time perception	Rhythmic Patterns and Rates of Activities tempo, time, patterns of durations.
Timbre (comprised of physical components: dynamic envelope, spectrum, and spectral envelope)	Timbre (perceived as overall quality; defined by the definition of fundamental frequency and by the analysis of the physical dimensions)	Sound Sources and Sound Quality sound sources, groupings of sound sources, instrumentation, performance intensity, texture (quality of the overall sound), performance techniques
Space (comprised of physical components created by the interaction of the sound source and the environment, and their relationship to a receptor)	Space (perception of the sound source as it interacts with the environment, and perception of the physical relationship of the sound source to the listener)	Spatial Properties lateral (stereo) location, imaging, distance, moving sources, depth of sound stage, phantom images, environmental characteristics, space within space

than the use of the other artistic elements. Complex relationships of pitch patterns and levels are quite common in music.

Information about the artistic element of pitch levels and relationships will be related to:

1. The relative dominance of certain pitch levels;
2. The relative registral placement of pitch levels and patterns; or

3. Pitch relationships: patterns of successive intervals, relationships of those patterns, and relationships of simultaneous intervals.

The artistic element of pitch levels and relationships is broken into the component parts: melodic lines, chords, tonal organization, register, range, textural density, pitch areas, and tonal speech inflection.

A series of successive, related pitches creates *melodic lines*. Melodic lines are perceived as a sequence of intervals that appear in a specific ordering and that have rhythmic characteristics. The melodic line is often the primary carrier of the artistic message of a piece of music.

The ordering of intervals, coupled with or independent from rhythm, creates patterns. *Pattern perception* is central to how humans perceive objects and events. These basic principles relate to all the components of the artistic elements. Melodic lines are organized by patterns of intervals (short melodic ideas, or motives) and supported by corresponding rhythmic patterns. The complexity of the patterns, the ways in which the patterns are repeated, and the ways in which the patterns are modified provide the melodic line with its unique character.

Two or more simultaneously sounding pitches create *chords*. In much music, these chords are based on superimposing, or stacking, the intervals of a third (intervals containing three and four semi-tones, most commonly). Chords comprised of three pitches, combining two intervals of a third, are called *triads*. The continued stacking of thirds results in seventh, ninth, eleventh, and thirteenth chords.

The movement from one chord to another, or *harmonic progression*, is the most stylized of all the components of the artistic elements. Harmonic progression is the pattern created by successive chords, as based on the lowest note (the root) of the triads (or more complex chords). These patterns of chord progressions have been established as having general principles that occur consistently in certain types of music. Certain types of music will have stylized chord progressions (progressions that occur most frequently), other types of music will have very different movement between chords and will perhaps emphasize more complex chord types. The patterns of the harmonic progression create *harmony*.

Harmony is one of primary components that supports the melodic line. Pitches of the melody are reinforced by the chords in the harmonic progression. The speed and direction of the melodic line is often supported by the speed at which chords are changed and the patterns created by the changing chords—*harmonic rhythm*.

The expectations of harmonic progression create a sequence of chords that will present areas of tension and areas of repose within the musical composition. The tendencies of *harmonic motion* do much to shape the momentum of a piece of music, and can greatly enhance the character of the melodic line and musical message. Performers utilize the psychological tendencies of harmonic progression, exploiting its directional and dramatic tendencies. The expectations of harmonic movement and the psychological characteristics of harmonic progression have become important aspects of musical expression and musical performance.

The melodic and harmonic pitch materials are related through *tonal organization*. Certain pitch materials are emphasized over others, to varying degrees, in nearly all music. This emphasis creates systems of tonal organization in which a hierarchy of pitch levels exist. A hierarchy will usually place one pitch in a predominant role, with all other pitches having functions of varying importance, in supporting the primary pitch. The primary pitch, or *tonal center*, becomes a reference level, to which all pitch material is related, and around which pitch patterns are organized.

Many tonal organization systems exist. These systems tend to vary significantly by cultures, with most cultures using several different, but related, systems. The "major" and "minor" tonal organization systems of Western music are examples of different, but related, systems, as are the "whole-tone" and "pentatonic" systems of Eastern Asia. The reader should consult appropriate music theory texts for more detailed information on tonal organization, as necessary.

Certain components of pitch levels and relationships have become more prominent in musical contexts (and other areas of audio) because of the artistic treatment that pitch relationships have received in music recordings. The components of range, register, textural density, and pitch area can be more closely controlled in recorded music than in live (unamplified) performance. These components are more important in recorded music because they are precisely controllable by the technology and have been controlled to support and enhance the musical material.

Range is the span of pitches of a sound source (or of any instrument or voice). Range encompasses the area of pitches from the highest note possible (or present in a certain musical example or context) to the lowest note possible (or present), in a particular sound source.

A *register* is a portion of a sound source's range. A register will have a unique character (such as a unique timbre or some other determining factor) that will differentiate it from all other areas of the same range. It is a small area within the source's range that is unique in some way. Ranges are often divided into many registers; registers may encompass a very small group of successive pitches, up to a considerable portion of the source's range.

A *pitch area* is a portion of any range (or register) that may or may not exhibit characteristics that are unique from other areas. Instead, it is a defined area between an upper and a lower pitch level, in which a specific activity or sound exists.

Textural density is the relative amount and registral placement of simultaneously sounding pitch material, throughout the hearing range or within a specific pitch area. It is the amount and placement of pitch material in the composite musical texture (the overall sound of the piece of music), in relation to defined boundaries.

With textural density, sound sources are assigned (or perceived as being within) a certain pitch area, within the entire audible range (or range used within a certain piece of music). Thus, certain pitch areas will have more activity than other pitch areas; certain sound sources will be present only in certain pitch areas, and other sources present only in other pitch areas; some sources may share pitch areas and cause more activity to be present in those portions of the range; some pitch areas may be void of activity. Many possible variations exist.

Textural density is a component of pitch level relationships that is directly related to traditional concerns of orchestration. Textural density is a much more specific concern in recorded music because it is controllable in very fine increments. Traditional orchestration was concerned, basically, with selecting instruments and placing the musical parts (performed by the assigned instruments) against one another.

With the controls of signal processing (especially equalization), sound synthesis, and multitrack recording, the registral placement of sound sources and their interaction with the other sound sources take on many more dimensions. Each sound source occupies a pitch area; the acoustic energy within the pitch area of a timbre's spectrum is distributed in ways that are unique to each sound source. The spectrum of each sound source is an individual textural density, and the textural density of the overall program (or musical texture) is the composite of all the simultaneous pitch information from all sound sources.

Sound sources and musical ideas are often delineated by the pitch area they occupy within the composite textural density. Sound sources are more easily perceived as being separate entities and individual ideas, when they occupy their own pitch area in the composite, textural density of the musical texture. This area can be large or quite small, and still be effective.

Sounds that do not have well-defined pitch quality, occupy a specific pitch area. These types of sound are noise-like, in that they cannot be perceived as being at a specific pitch. Such sounds do, however, have unique pitch characteristics.

Many sounds cannot be assigned a specific pitch, yet have a number of frequencies that dominate their spectrum. Cymbals and drums easily fall into this category. Cymbals are easily perceived as sounding higher or lower than one another. Yet a specific pitch cannot be assigned to the sound source.

These sounds are perceived as occupying a pitch area. A pitch-type quality is perceived in relation to the registral placement of the area in which the highest concentration of pitch information (at the highest amplitude level) is present in the sound, and in relation to the relative density (closeness of the spacing of pitch levels) of the pitch information (spectral components). One is able to identify the approximate area of pitches in which the concentration of spectral energy occurs, and is thus able to relate that area to other sounds.

Pitch areas are defined as the range spanned by the lowest and highest dominant frequencies around the area of the spectral activity. This range is called the *bandwidth* of the pitch area. Many sounds will have several pitch areas, where concentrated amounts of spectral energy is occurring, with one range dominating and others less prominent. The size of the bandwidth and the density of spectral information (the number of frequencies within the bandwidth and the spacing of those frequencies) define the sound quality of the pitch area.

DYNAMIC LEVELS AND RELATIONSHIPS

Dynamic levels and relationships have traditionally been used in musical contexts for expressive or dramatic purposes. Expressive changes in dynamic levels and the relationships of those changes have most often been used to support the motion of melodic lines, enhance the sense of direction in harmonic motion, or emphasize a particular musical idea. A change of dynamic level, in and of itself, can produce a dramatic musical event and is a common musical occurrence. Changes in dynamic level can be gradual or sudden, subtle or extreme.

Dynamics have traditionally been described by analogy: louder than, softer than, very loud (fortissimo), soft (piano), medium loud (mezzoforte), and so forth. The artistic element of dynamics in a piece of music is judged in relation to context. Dynamic levels are gauged in relation to (1) the overall, conceptual dynamic level of the piece of music, (2) the sounds occurring simultaneously with a sound source in question, and (3) the sounds that immediately follow and precede a particular sound source.

The components of dynamic levels and relationships in audio recording are dynamic contour (with gradual and abrupt changes in dynamic level), emphasis/deemphasis accents (abrupt changes in dynamic level), musical balance (gradual and abrupt changes in dynamic levels), and dynamic speech inflections.

Rapid, slight alterations or changes in dynamic level for expressive purposes are often present in live performances. This is called *tremolo* and is used primarily to add interest and substance to a sustained sound. Tremolo and vibrato are often confused. *Vibrato* is a rapid, slight variation of the pitch of a sound; it is also used to enhance the sound quality of the sound source. At times, performers may not be able to control their sound well enough to control tremolo and vibrato alterations; in these instances, tremolo and vibrato may detract from the source's sound quality, rather than contribute to it.

Changes in dynamic levels over time comprise *dynamic contour*. Dynamic contours can be perceived for individual sounds, individual sound sources, individual musical ideas comprised of a number of sound sources, and the overall piece of music. Dynamic contour can be perceived from many different perspectives—from the smallest changes within the spectral envelope through great changes in the overall dynamic level of a recording.

The composite of all of the dynamic contours creates *musical balance*. Musical balance is the interrelationships of the dynamic levels of each sound source to one another and to the entire musical texture. The relative dynamic level of a particular sound source in relation to another sound source is a comparison of two parts of the musical balance.

Dynamic contours and musical balance have been used in supportive roles in most traditional music. At times, dynamic level changes have been used for their own dramatic impact on the music, but most often they are used to assist the effectiveness of another artistic element.

To support a musical idea or to create a sense of drama, musical ideas are often brought to the listener's attention by dynamic *emphasis* or *attenuation accents*. An accent is a shift in dynamic level that brings the listener's attention to a musical idea. Accents are usually emphasis accents, increasing the dynamic level of the sound, to achieve the desired result. Much more difficult to successfully achieve, de-emphasis (or attenuation) accents draw the listener's attention to a musical idea, or a sound source, by decreasing the dynamic level of the sound. Attenuation accents are often unsuccessful because the listener has a natural tendency to move attention away from softer sounds; these accents are most easily accomplished in sparse musical textures, where little else is going on that will draw the listener's attention away from the material being accented.

Dynamic levels and relationships may be significantly different in the final recording than they were in the original performance. The recording process has very precise control over the dynamic levels of a sound source in the musical balance of the final recording. An instrument may have an audible dynamic level in the musical balance of a recording that is very different from the dynamic level at which the instrument was originally performed. The timbre of the instrument will exhibit the dynamic levels at which it was performed (*performance intensity*), but its relative dynamic level in relation to the other musical parts might be significantly altered by the mix. For example, an instrument may be recorded playing a passage at *ff*, with the passage ending up in the final musical balance at a very soft dynamic level. The timbre of the instrument will send the cue that the passage was performed very loudly, yet the actual dynamic level will be quite soft in relation to the overall musical texture and to the other instruments of the texture.

The dynamic level of a sound source in relation to other sound sources—musical balance—is quite different and distinct from the perceived distance of one sound source to another. Yet these two occurrences are often confused and are the source of much common, misleading terminology used by recordists. Significant differences are present between a softly generated sound that is close to the listener, and a loudly performed sound that is at a great distance to the listener, even when the two sounds have precisely the same perceived loudness level. Loudness levels within the recording process are independently controllable from the loudness level at which the sound was performed and from the distance of the sound source from the original receptor and from the person listening to the final recording.

RHYTHMIC PATTERNS AND RATES OF ACTIVITIES

Durations of sounds (the length of time in which the sound exists) combine to create musical rhythm. Rhythm is based on the perception of a steadily recurring, underlying pulse. The pulse does not need to be strongly audible to be perceived. The underlying pulse (or metric grid) is easily recognized by humans as the strongest, common proportion of duration (note value) heard in the music.

The rate of the pulses of the metric grid is the *tempo* of a piece of music. Tempo is measured in metronomic markings (pulses per minute, abbreviated "M.M.") or, in some contexts, as pulses per quarter note. Tempo, in a larger sense, can be the rate of activity of any large or small aspect of the piece of music (or of some other aspect of audio—for example, the "tempo of a dialogue").

Durations of sound are perceived proportionally in relation to the pulse of the metric grid. The human mind will organize durations into groups of durations, or *rhythmic patterns*. In the same ways that people perceive patterns of pitches, they perceive patterns of durations. Pattern perception is transferable to all the components of all the artistic elements, and is the traditional way in which pitch and rhythmic relationships are perceived.

Rhythmic patterns are the durations of or between occurrences of an artistic element. Rhythmic patterns might be created by the pulsing of a single percussion sound; in this way, rhythmic patterns would be created by the durations between the occurrences of the starts of the same sound source. Rhythmic patterns comprised of the durations of successive, single pitches (perhaps including some silences) creates melody. Rhythmic patterns of the durations of successive chords (groups of pitches) creates harmonic rhythm. In this way, rhythm can be transferred to *all* artistic elements. For example, it is possible to have rhythms of sound location (as is becoming a very common mixing technique for percussion sounds); it is likewise possible to have timbre melodies, which are rhythms applied to patterns of identifiable timbres.

SOUND SOURCES AND SOUND QUALITY

Selecting, modifying, or creating *sound sources* is an important artistic element of audio recording. The *sound quality* of the sound sources (the timbre of the source) plays a central role in presenting musical ideas, and has become an increasingly significant resource for expressing musical ideas.

The sound quality of a sound source may cause a musical part to stand out from others or to blend into an ensemble. In and of itself, it can convey tension or repose, or lend direction to a musical idea. It can also add dramatic or extra-musical meaning or significance to a musical idea. Finally, the sound quality of a sound source can, itself, be a primary musical idea, capable of conveying a meaningful musical message.

Until the twentieth century, composers of Western music used the sound quality of a sound source:

1. To assist in delineating and differentiating musical ideas;
2. To enhance the expression of a musical idea by carefully selecting the appropriate musical instrument to perform a particular musical idea; or
3. To create a composite timbre (or *texture*) of the ensemble, thereby forming a characteristic, overall sound quality.

Performers have always utilized the characteristic timbres of their instruments or voices, to enhance musical interpretation. This activity has been greatly refined by the resources of recording technology. The recording process allows performers greater flexibility in shaping the timbre of their instruments, for creative expression. Of equally great importance, after the performance has been captured, the recording process provides the opportunity to return to the performance for further (perhaps extensive) modifications of sound quality.

Selecting a sound source to represent (present) a particular musical idea is vital to the successful communication of the idea. The act of selecting a sound source is among the most important decisions composers (and producers) make. The options for selecting sound sources are: (1) to choose a particular instrumentation, (2) modifying the sound quality of an existing instrument or performance, or (3) to create, or synthesize, a sound source to meet the specific needs of the musical idea.

The *selection of instrumentation* was once a matter of merely deciding which generic instrument of those available would perform a certain musical line. The selection of instrumentation has become very specific, since the performance of a recording may virtually live forever, whereas previous performances existed for only a passing moment.

Today, the selection of instrumentation is often so specific, as to be a selection of a particular performer playing a particular model of an instrument. Generally, composers and producers are very much aware of the sound quality they want for a particular musical idea. The performer, the way the performer can develop a musical idea through personal performance techniques and his or her ability to use sound quality for musical expression are all considerations in the selection of instrumentation.

Vocalists are commonly sought for the sound quality of their voice, as well as their abilities to perform in particular singing styles. The vocal line of most songs is the focal point that carries the weight of musical expression. Vocalists make great use of performance techniques to enhance and develop their sound quality, as well as to support the drama and meaning of the text.

Performance techniques vary greatly between instruments, musical styles, performers, and functions of a musical idea. The most suitable performance techniques will be those that achieve the desired musical results, when the sound sources are finally combined. One performance technique consideration must be singled out for special attention: the intensity level of a performance.

A performance on a musical instrument will take place at a particular intensity level. This *performance intensity* information is comprised of loudness, performance technique, and the expressive qualities of the performance. Each performance at a different intensity level results in a characteristic timbre of that instrument, at that loudness level. The same sound source will have different timbres, at different loudness levels (and at different pitch levels), and so on, through performance intensity.

Along with the timbre (sound quality) and the loudness level comes a sense of drama and an artistically sensitive presentation of the music that is communicated to the listener. Through performance intensity, louder sounds might be more urgent, more intense, and softer sounds might be cause for relaxing musical motion. Much dramatic impact can be created by sending conflicting loudness level and sound quality information—a loud whisper, a trumpet blast heard at pianissimo.

Modifying an existing sound source is a common way of creating a desired sound quality. Instruments, voices, or any other sound may be modified (while being recorded, or afterwards), to achieve a desired sound quality. Most often, this option for selecting a sound source is in the form of making detailed modifications to a recorded performance of a musical idea by a particular instrument. The final sound quality will still have the characteristic qualities of the original sound.

Extensively modifying an existing sound source, to the point where the characteristic qualities of the original sound are lost, is actually the *creation of a sound source*. Creating new sound qualities (or inventing timbres) has become an important feature in many types of music. The recording process easily allows for the creation of new sound sources, with new sound qualities.

Sound qualities are created by either extensively modifying an existing sound (through sound sampling technologies) or by synthesizing a waveform. Sound synthesis techniques allow precise control over these two processes, and are having a widespread impact on recording practice and musical styles. Many specific techniques exist for synthesizing and sampling sounds, all with their own unique sound qualities and own unique ways of allowing the user to modify the sound source.

With the control of timbre by the recording process has come a new sense of the importance of sound quality to communicate, as well as to enhance, the musical message. Sound quality has become a central element in a number of the primary decisions on recording music and in creating music through the recording process. In making these primary decisions, sound quality is conceptualized as an object. The sound is thought of as a complete and individual entity.

In this way, sound quality is considered as a *sound object*. While the sound object is comprised of component parts (as discussed earlier), it is perceived as a large unit, for its overall sound qualities.

Sound quality is perceived as a sound object when (1) the sound quality of the sound source itself is at the center of the listener's attention, or (2) the sound itself is the most important element of the musical texture.

For the sound object, the individual character of the sound source is significant. This is in contrast to the normal, primary significance of how the sound quality enhances the musical material, or how the sound sources relate to one another.

The entire sound of the music may also be conceptualized as a single entity, or overall sound quality. In this way, the overall musical sound is perceived as a large sound object, being comprised of any number of small, individualized sound objects.

This sound quality of the overall sound, or entire program, is texture. As texture is the composite of all sound objects present at any one time, or over a span of time, it has also been called *sound structure* or *sound event*.

Texture is perceived by the characteristics of its global sound quality. This concept of sound quality can be applied to groups of sound sources, in the same way as to individual sound objects or the entire program.

Texture will nearly always be comprised of any number or types of individual sound objects or groups of sound sources. Texture is perceived as an overall character that is comprised of the states and activities of its component parts. Pitch-register placements, rate of activities, dynamic contours, and spatial properties are primary factors in defining a texture by the states or values of its component parts.

SPATIAL PROPERTIES

The *spatial properties* of sound have traditionally not been used in musical contexts. The only exceptions are the location effects of antiphonal ensembles of certain Renaissance and early twentieth-century musics, and the effect of the movement of the sound source found in certain drama-related works of the nineteenth century.

The spatial properties of sound play an important role in communicating the artistic message of recorded music. The roles of spatial properties of sound are many: it may be to enhance the effectiveness of a large or small musical idea; it may help to differentiate one sound source from another; it may be used for dramatic impact; it may be used to alter reality or to reinforce reality.

The number and types of roles that spatial location may play in communicating a musical idea have yet to be exhausted or defined.

All the components of the spatial properties are under very precise and independent control. All the spatial properties have the capacity to be in, and to gradually change between, many dramatically different and fully audible states.

The spatial properties of sound that are of primary concern to recorded music (sound) are (1) the perceived *stereo location* of the sound source on the horizontal plane of the stereo array, (2) the conceptualized *distance* of the sound source from the listener, and (3) the perceived characteristics of the sound source's physical *environment*. The perceived elevation of a sound source is not consistently reproducible in widely used playback systems, and has not yet become a resource for artistic expression.

The three spatial properties are realized through stereophonic sound reproduction. The spatial attributes are related by the perceived relationships of location and distance cues of the sound sources in relation to the *sound stage*, and the relationships of the sound stage to the *perceived performance environment* of the recording.

Two-channel sound reproduction has become the standard for the recording industry, with monophonic reproduction still used in AM broadcast and most television sound. The two-channel array of *stereo sound* attempts to reproduce all spatial cues through two separate sound locations (loudspeakers), each with more-or-less independent content (channel). With the two channels, it is possible to create the illusion of sound location at a loudspeaker, in between the two loudspeakers, or slightly outside the boundaries of the loudspeaker array. Location is limited to the area slightly beyond that covered by the stereo array, and to the horizontal plane. The characteristics of the sound source's environment and distance from the listener are affected in much more subtle ways by the stereo reproduction format.

A setting is created by the two-channel playback format, for recreating a recorded or created performance (complete with spatial cues). The setting of the two-channel playback format is a conceptual (and physical) environment within which the recording will be reproduced more-or-less accurately.

The reproduced recording presents an illusion of a live performance. This performance will be perceived as having existed in reality, in a real physical space, as the human mind will conceive of any human activity in relationship to their own physical experiences. The recording will appear to be contained within a single, perceived physical environment.

The sound stage is the location within the perceived performance environment, where the sound sources appear to be sounding. The sound stage will appear to be contained within a single, global environment. The sound sources of the recording will be grouped by the mind, and will appear to occupy a more or less specific area of that global environment; this area is the sound stage. It is possible for different sound sources to occupy significantly different locations within the sound stage.

Imaging is the perceived lateral location and distance placement of the individual sound sources within the sound stage. Imaging is defined by the perceived physical relationships of the sound sources. As such, it is the perceived locations of the sound sources within the stereo array and with respect to perceived distance.

The stereo (lateral) location of a sound source is the perceived placement of the sound source, in relation to the stereo array. Sound sources may be perceived at any lateral location within, or slightly beyond, the stereo array.

Phantom images are sound sources that are perceived to be sounding at locations where a physical sound source does not exist. Imaging relies on phantom images to create lateral localization cues for sound sources. Through using phantom images, sound sources may be perceived at any physical location within the stereo loudspeaker array, and up to 15 degrees beyond the loudspeaker array. *Stage width* (sometimes called *stereo spread*) is the area that spans the boundaries established by extreme left and right images of the sound sources. Phantom images not only provide the illusion of the location of a sound source, but also create the illusion of the physical size of the source. Two types of phantom images exist: the *spread image* and the *point source*.

The point source phantom image is a focused, precise point in the sound stage, where a sound source appears to be located. It is an exact location that the sound source is perceived to be occupying. It appears to have a physical size that is quite narrow and precisely located in the sound stage.

The spread image appears to occupy an area. It is a phantom image that extends between two audible boundaries. The size of the spread image can be considerable; it might be slightly wider than a point source, or it may occupy the entire stereo array. The spread image is defined by its boundaries; it appears to occupy an area between two points. At times, a spread image may appear to have a "hole in the middle," where it might occupy two equal areas, one on either side on the stereo array.

The perceived lateral location of sound sources can be altered to provide the illusion of *moving sources*. Moving sound sources may be either point sources or spread images. Point sources that change location resemble most closely one's experiences of moving sound sources.

The listener will perceive two types of *distance cues* from the recorded music: (1) the distance of the listener to the sound stage, and (2) the distance of each sound source from the listener. Both of these distances rely on a perception that the entire recording occupies a single, global environment. This perceived performance environment establishes a reference location of the listener, from which all judgements of distance can be calculated.

The *stage-to-listener distance* establishes the location of the sound stage with respect to the listener. It is the distance between the grouped sources that comprise the sound stage and the audience. This stage-to-listener distance is the placement of the sound stage within the overall environment of the recording, in relation to the perceived location of the listener.

The *depth of sound stage* is an area occupied by the distance of each sound source, relative to one another. The boundaries of the depth of the sound stage are the perceived nearest and furthest sound sources. The perceived distances of sound sources within the sound stage may be extreme.

The two factors of distance cues interact. The depth of the sound stage is perceived in the context of the stage-to-listener distance; the listener is prone to placing the nearest sound source at the nearest location of the stage-to-listener distance. Conversely, the perceived distance of each sound source, relative to the listener, can cause a shift in the perceived stage-to-listener distance, especially in multitrack recordings that incorporate dramatic reverberation techniques.

These two factors of distance cues have different levels of importance in different contexts. Depth of sound stage cues tend to be emphasized over stage-to-listener distance cues in many recordings; the cues of the distance of the source from the listener are often exploited, to support dramatic and musical ideas. As another example, stage-to-listener distance cues are carefully calculated in many art music recordings (especially those utilizing standardized stereo microphone techniques). While the distance might not change within the recording, the stage-to-listener distance is carefully selected, to represent the most appropriate vantage point (the ideal seat) from which to hear the music.

Matching a sound source to the sound characteristics of an environment in which it will sound, and selecting the environment of the sound stage (the perceived performance environment) have become important parts of music recording. This coupling of source to *environmental characteristics* can:

- Have a significant impact on the meaning of the music, of the text, or of the sound source;
- Supply dramatic effect;
- Segregate sound sources, musical ideas, or groups of instruments; and
- Enhance the function and effectiveness of a musical idea.

The sound characteristics of the host environments of sound sources and the complete sound stage are precisely controllable. Each sound source has the potential to be assigned environmental characteristics that are different from the other sound sources. The recording process allows for different environments to be assigned to different sound sources, and for those characteristics to be varied, as desired. Further, each source may occupy any distance from the listener, within the applied host environment.

The environment of the sound stage and the individual environments for each sound source (or groups of sound sources) often coexist in the same music recording. This musical context places the individual sound sources with their individual environments "within" the overall environment of the recording. The result is a perception that:

1. Physical spaces may exist side by side;
2. One physical space may exist within another physical space (to the point where a larger physical space may be perceived to exist within a smaller physical space); and
3. Sounds may exist at various distances within the same or different host environments, within other environments (causing conflicting distance cues between sources and environments).

The result is the illusion of *space within space.*

Any number of environments and associated stage-depth distance cues may occur simultaneously and coexist within the same sound stage. The environments and associated distances are conceptually bound by the spatial impression of the perceived performance environment. These "outer walls" of the overall program establish a reference (subconsciously, if not aurally) for comparing the sound sources. Oddly, the overall space that serves as a reference, and that is perceived by the listener as being the space within which the other activities occur, might have the sound characteristics of a physical environment that is significantly smaller than the spaces of the sound sources it appears to contain. Such cues that send conflicting messages between life experiences and the perceived musical occurrence can be used to great artistic advantage. This is a very common space within space relationship.

Space within space will at times be coupled with distance cues to accentuate the different environments (spaces) of the sound sources. Often, this illusion will be created solely by the environmental characteristics of the different spaces of each sound source.

Space within space has become an important element in shaping the imaging of a recording. Often, imaging will work in a complementary and contrasting fashion with musical balance. The degree of interaction of these two artistic elements is often quite sophisticated in modern music recordings.

With the recording process, it is possible for any of the artistic elements of sound to be varied in considerable detail. In so doing, all artistic elements can be shaped for artistic purposes, or used as resources for artistic expression. As all elements of sound are capable of an equal amount of variation, it is possible for each element of sound to function in any role in communicating the message of a piece of music.

The artistic elements are used in very traditional roles in certain musical works and types of recording productions, and in very new ways in other works. The new ways that the artistic elements are used tend to place more emphasis on the artistic elements of sound that cannot be precisely controlled acoustically. Many new musical compositions use the artistic elements unique to audio recording (especially sound quality and spatial properties), to support the musical material. Different musical relationships and sound properties exist in audio recordings than can be found in the music conceived before the artistic resources of recording technology existed.

The potentials of the artistic elements to convey the musical message, the musical message itself, and the characteristics and limitations of the listener are explored in the next chapter.

3

The Musical Message and the Listener

This chapter will discuss the musical message, how the artistic elements are perceived in the contexts of audio production and music, and how the elements function in communicating the message of a piece of music. The perception of the artistic meaning of the music is largely determined by the listener's ability to correctly interpret the sounds of the musical message. The factors that limit the listener's ability to effectively interpret the artistic elements of sound into the intended musical message (or meaning of the music) will be explored. Also the "listener" as audience member, and as audio professional, will be contrasted.

CRITICAL LISTENING VERSUS ANALYTICAL LISTENING

The artistic elements are the functions of the physical dimensions of sound, applied to the art of recording. People judge the physical dimensions of sound through their perception, as perceived parameters. Judgements related to the integrity of sound quality are contrasted with judgements related to the meaning of the states of the artistic elements.

The same aspects of sound quality may provide two different sets of information. This is entirely dependent upon the way one listens to the sound material: evaluating the sound for its own content (critical listening), or evaluating sound for its relationships to context (analytical listening).

The recordist must understand how the components of sound function in relation to the various aspects of a piece of music and in relation to understanding the message of the piece itself (analytical listening), as well as how the components of sound function to create the impression of a single sound quality and in relation to the technical quality of the audio signal (critical listening). Most methods of listening, for the audio professional, will be much more detailed and specific than those required of the lay listener.

The people listening to a final piece of music do not need to be aware of the small details of the quality of sound that go into a well-produced recording. This information (related to critical listening) is simply not necessary in order to appreciate the artistic message (or the emotive experience) of the music, which is, after all, the purpose for their listening experience. The lay audience of the recording will not be consciously aware that certain artistic elements are presenting important information needed to understand a piece of music, and the audience will not have the knowledge and experience to accurately identify and define the information (related to analytical listening).

Both of these processes are required of the audio professional, but not of the lay listener. The shaping of sound quality and the refinement of musical ideas that go into the recording may enhance the audience's listening experience, but in all likelihood will not be consciously perceived by the lay audience. These two areas are central to the concerns of the recordist, however, and require two different ways of listening.

All listeners must contend with a set of limitations that relate to their own experience and knowledge. For the audience, these limitations will inhibit the understanding of the piece of music. For the audio professional, these limitations will restrict the ability of the individual to perform meaningful evaluations of sound—a process that is central to nearly all positions in the industry.

Audio professionals evaluate sound in two ways: critical listening and analytical listening.

Analytical listening is the evaluation of the content and the function of the sound, in relation to the musical or communication context in which it exists. Analytical listening is the definition of the function (or significance) of the musical material (or sound) to the other musical materials in the structural hierarchy. This type of listening is a detailed observation of the interrelationships of all musical materials and of any text (lyrics). It will enhance the recordist's understanding of the music being recorded, and will allow the recordist to conceive of the artistic elements as musical materials that interact with traditional aspects of music.

Critical listening is the evaluation of the characteristics of the sound itself, and/or of the integrity of the audio signal (technical quality) through human perception (or the evaluation of sound quality out of context). Critical listening is the process of listening to the *dimensions* of the artistic elements of sound in a new application—out of the context of the music. In critical listening, the states and values of the artistic elements function as subparts of the perceived parameters of sound. These aspects of sound are perceived in relation to their contribution to the characteristics of the sound, or "sound quality."

Critical listening seeks to define the perception of the physical dimensions of sound, as they are exhibited by the recording process. It is concerned with making evaluations of the characteristics of the sound itself, without relation to the material surrounding the sound, or the meaning of the sound. Critical listening must take place at all levels of listening *perspective* (see a following section), from the overall program to the most minute aspects of sound.

THE MUSICAL MESSAGE

The message of a piece of music is determined by the many purposes or functions of music. There are different ways to listen to music, and different purposes for music to serve. People listen to music with various levels of attention; at the extremes, their minds will be focused and intent on extracting certain specific types of information, or their attention will be focused on some activity other than the music, not being conscious of the music.

The purpose of a piece of music, and thereby the characteristics of the musical message, may take the forms of (1) conceptual communication, (2) portraying an emotive state, (3) aesthetic experience, and (4) utilitarian functions. The purposes are by no means exclusive; many pieces of music use different functions simultaneously, or at different points in time.

Music that includes a text, such as a song, will communicate other concepts. These works may tell a story, deliver the author's impressions of an experience, present a social commentary, and so forth. Music is used as a vehicle to deliver the tangible ideas of the author/composer. The interplay between the music and the drama of the text is often an important contributor to the total experience of these works.

It is difficult for music alone to communicate specific concepts. Works are often written that portray certain subjects without a text, by shaping the music to cause the listener to associate the sounds of the music to his or her experiences of the subject. The subjects of such works are often general in nature.

Certain concepts are associated with certain specific or types of musical materials (types of movie music, musical ideas associated with certain individuals or certain landscapes, etc.). These are exceptional cases where music alone can communicate specific concepts, with the aid of associations drawn from the listener's past experiences.

Music communicates emotive states easily. Music may portray a specific mood or incite a specific emotional response from the listener, or it may create a more general and hard to define (yet convincing) feeling or emotive impression. The composer draws from the past experiences of the listener, to shape emotive reactions to the material.

Music may be an aesthetic experience. The perception of the relationships of the musical materials alone, without the associations of concepts or emotive states, may be the vehicle for the musical message. Music has the ability to communicate on a level that is separate and distinct from the verbal (conceptual) or the emotional. In this way, music (as all of the arts) can reach beyond the human experience; ideas that cannot be verbally defined or represented emotively are communicated. Abstract concepts may be clearly communicated; the human spirit may reach beyond reality, to loftier ideals. The experience of the aesthetic appreciation of art has been compared to the impressions of religious experiences, by some people.

Music also serves other functions. It is used to reinforce or accompany other art forms (motion pictures, musical theater, dance, video art), to enhance the audio and visual media (television, radio, advertisements), and to fill "dead air" in everyday experiences (supermarkets, elevators, offices). In these instances, music is present to support dramatic or conceptual materials, to take the listener's attention away from some other sounds or activities (a dentist's office), or to make an environment more desirable (a restaurant, an automobile).

The complexity of the musical materials is often a direct reflection of the function of the music. When music is the most important aspect of the listener's experience, the musical material may be more complex, since the listener will devote more effort to deciphering the materials. When the music is playing a supportive role, the materials are often less sophisticated and are directly related to the primary aspect of the listener's experience. When music is being used to cover undesirable noises or to fill a void of silence that would otherwise be ignored, the musical materials are often very simple, easily recognized, and easily "heard," without requiring the listener's attention.

MUSICAL FORM AND STRUCTURE

Within the human experiences of time and space, nothing exists without shape or form. Music is no exception. Pieces of music have *form* as a global quality, as an overall conception.

Pieces of music can be conceptualized as an overall quality. It is the human perception of form that provides the impression of a global quality and crystallizes the entire work into a single entity. Form is the piece of music as if perceived, in its entirety, in an instant–the shape or design that is perceived from conceptualizing the whole.

Form is the global shape or design of a piece of music that is the sum that is shaped from the interactions of its component parts. Form is comprised of component parts that are the materials of the piece of music. The materials of the music and their interrelationships provide the *structure* of the work.

The structure of a work is the architecture of the musical materials. Structure is the characteristics of the musical materials coupled with a *hierarchy* of the interrelationships of the musical materials, as they function to shape the work. The artistic elements of sound function to provide the musical materials with a unique character, as will be discussed below; the musical materials are related by structure.

The hierarchy of musical materials is the listener's perception of a general framework of the materials. It is an interrelationship of the materials (with their varying levels of importance) to one another, and to the musical message as a whole. Within the hierarchy of musical structure, all musical materials and artistic elements will have a greater importance to the musical message than other materials or elements (except the least significant), all sections or ideas in the music will have greater importance over others (except the least significant), and all musical materials are subparts of other, more significant musical materials (except the most significant materials).

Further, the hierarchy of the musical structure organizes musical materials into patterns, and patterns of patterns. In this way, relationships are established between the subparts of a work and the work as a whole (Fig. 3-1). The hierarchy is such that any time-span may contain any number of smaller time-spans, or may be contained within any number of larger time-spans; musical material at any level may be related to material at any other level of the hierarchy.

A multitude of possibilities exist for unique musical structures, supporting the same musical form. The multitude of popular songs that have appeared on the Top-40 charts during the past few decades is evidence of this great potential for variation. Most of the songs have many similarities in their structures, but they also have many significant differences. The materials that comprise the works may be very different, but the materials work towards establishing an overall shape (form) that is quite similar between songs.

Many songs share the same or similar forms. Their overall conceptions are very similar, although the materials of the music and the interrelationships of those materials may be strikingly different. Form is an overall design that may be constructed of a multitude of materials and relationships.

Musical materials can be changed to dramatically alter the structure of a piece of music without altering its form. Many different structures can lead to the same overall design (form).

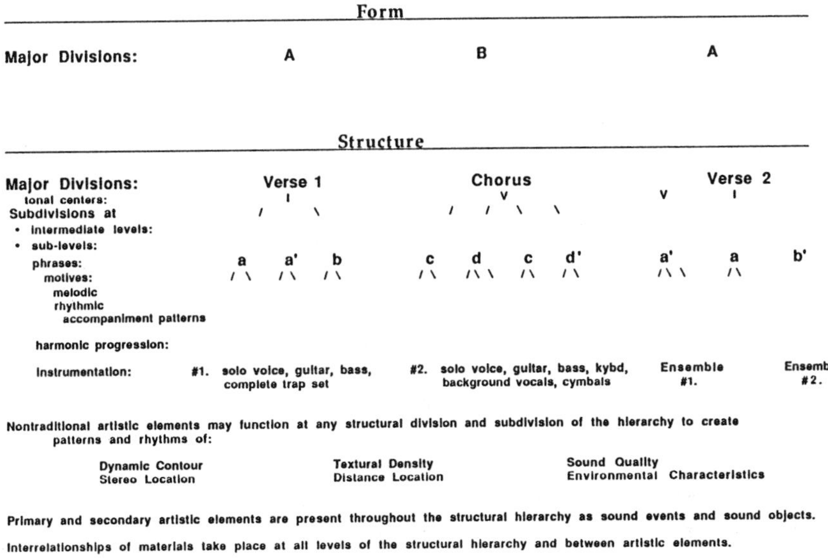

Form

Major Divisions: A B A

Structure

Major Divisions:	Verse 1			Chorus				Verse 2		
tonal centers:	I			V				V	I	
Subdivisions at • intermediate levels: • sub-levels:	/ \			/ / \ \						
phrases:	a	a'	b	c	d	c	d'	a'	a	b'
motives: melodic rhythmic accompaniment patterns	/ \	/ \	/ \	/ \	/ \ \	/ \	/ \	/ \ \	/ \	
harmonic progression:										
instrumentation:	#1. solo voice, guitar, bass, complete trap set			#2. solo voice, guitar, bass, kybd, background vocals, cymbals				Ensemble #1.		Ensemble #2.

Nontraditional artistic elements may function at any structural division and subdivision of the hierarchy to create patterns and rhythms of:

Dynamic Contour, Stereo Location, Textural Density, Distance Location, Sound Quality, Environmental Characteristics

Primary and secondary artistic elements are present throughout the structural hierarchy as sound events and sound objects.

Interrelationships of materials take place at all levels of the structural hierarchy and between artistic elements.

FIGURE 3-1. Form and Structure in Music.

MUSICAL MATERIALS

As music moves and unfolds through time, the mind grasps the musical message through understanding the meaning and significance of the progression of sounds. During this progression of sounds, the mind is drawn to certain artistic elements (that create the characteristics of the musical materials). It perceives the *musical materials* as small patterns (small musical ideas, often called motives or gestures), and groups these small patterns into related larger patterns. The listener remembers the patterns, together with their associations to larger and smaller patterns (perceiving the structural hierarchy). In order for the listener to remember patterns, the listener must recognize some aspect of the organization of a pattern or some of the materials that comprised a pattern(s).

Contrasting, repeating, and varying patterns throughout the structure creates logic and coherence in the music. Several general ways in which musical materials are used and developed should make clear the multitude of possibilities. Materials are contrasted with other materials at the same and at different hierarchical levels (above and below). Materials are repeated immediately, or later in time, at the same or at different hierarchical levels. Materials are varied by adding or deleting portions of the idea, by altering a portion of the idea, or perhaps by transposing the idea to different artistic elements (such as melodic ideas becoming rhythmic ideas or harmonic motion becoming dynamic motion).

A balance of similarities and differences within and between the musical materials are required for successfully engaging music. A musical work will not communicate the desired message if a balance is not perceived by the listener.

The listener remembers the context in which the patterns were presented, as well as the patterns themselves. In presenting materials, some patterns will draw the listener's attention and be perceived as being more important than other patterns–these are the *primary musical materials*. Other patterns will be perceived as being subordinate. These *secondary materials* will somehow enhance the presentation of the primary materials by their presence and activity in the music. The secondary materials that accompanied the patterns (or primary materials) are also remembered, as individual entities (capable of being recognized without the primary musical idea) and as being associated with the particular musical idea (patterns).

The primary materials are traditionally melody (with related melodic fragments or motives) and (extra-musically) any text or lyrics of the music. The secondary materials are traditionally accompaniment passages, bass lines, percussion rhythms, harmonic progressions, and tonal centers. Secondary materials may also be dynamic contour, textural density, timbre development, stereo location, distance location, or environmental characteristics.

The secondary materials usually function to support the primary musical ideas. It is possible to have any number of equal, primary musical ideas. The potential groupings of primary and secondary musical ideas, in creating a single structural hierarchy, are limitless. Any number of secondary ideas (of varying degrees of importance in their support of the primary musical idea or ideas) may coexist in a musical texture, with any number of related or unrelated primary musical ideas.

Musical materials are given their unique characters by the states and values of the artistic elements of sound. The artistic elements of sound function to shape and define the musical materials.

THE RELATIONSHIPS OF ARTISTIC ELEMENTS AND MUSICAL MATERIALS

Musical ideas are comprised of *primary elements* and *secondary elements*. The primary elements are those aspects of sound that directly contribute to the basic shape or characteristics of a musical idea. The secondary elements are those aspects of the sound that assist the primary elements.

It is possible (in fact, common) to have more than one primary element and more than one secondary element contributing to the basic character of a musical idea. Primary elements exhibit changes in states and values that provide the most significant characteristics of the musical material. The secondary elements provide support in defining or in providing movement to the primary elements.

At all levels of the structural hierarchy, musical materials (primary and secondary musical ideas) are comprised of primary and secondary artistic elements. Therefore, it is possible for a certain element of sound to be a primary element on one level of the hierarchy and a secondary element on another level. This is not an uncommon situation. For example, a change in dynamic level of a drum roll might have primary significance at the hierarchical level of the individual sound source. At the same time, but at the hierarchical level of the composite sound of the entire ensemble, changes in dynamics are insignificant to the communication of the musical message, with pitch changes being of primary importance.

All the artistic elements of sound have the potential to function as the primary elements of the musical material; they have the potential to be the central carriers of the musical idea. Likewise, all the artistic elements of sound have the potential to function as secondary elements of the musical material, as well as in supportive roles, in relation to conveying the musical message. This will be thoroughly explored in the following pages.

In much current music, pitch remains the central element, or the primary carrier of the musical message. The new potentials of audio recording are being put into practice in the supportive roles of music, much more than as the primary elements. Most often, current production practice will utilize the new artistic elements, such as the stereo location of a sound source, to support the primary message (or to assist in defining the individual sound sources); rarely do the new sound resources function as the primary element of the primary musical idea.

The new musical resources can adequately function as the primary carriers of the musical material. Current practice is likely to continue its gradual change towards further emphasis of these new artistic elements of sound. It is important to recognize that the potential exists for any artistic element of sound to be the primary carrier of the musical material. The potential exists for any of the artistic elements of sound to function in support of any component of the musical idea. All the artistic elements are equally capable of change, and that change can be perceived almost equally well in all of the artistic elements.

Traditionally, a breakdown of primary and secondary elements of a piece of music (with associated musical materials identified) would commonly appear, as in the following table. Pitch is the primary element and is supported by rhythm and dynamics; the musical materials are differentiated by timbre differences.

Primary Elements	*Secondary Elements*
Pitch–melodic line #1	Pitch–harmony
Pitch–melodic line #2	Pitch–accompaniment patterns
	Dynamics–contour for expression
	Rhythm–supporting melody
	Timbre–instrument selection

With the potentials of audio recording, a similar (and equally common) outline might appear, as in the next table. While pitch remains a primary element, rhythm and timbre changes are of equal importance to delivering the musical message (timbre changes, in particular, is a new musical resource afforded by recording technology). Pitch, dynamics, and rhythm still play supportive roles, with spatial properties assisting timbre in differentiating the musical ideas.

Primary Elements	*Secondary Elements*
Pitch–melodic line #1	Pitch–harmony
Pitch–melodic line #2	Pitch–accompaniment patterns
Rhythm–recurring patterns	Dynamics–contour changes without associated changes in timbre; accents; contour for expression
Timbre–changing texture	Rhythm–supporting melody
	Timbre–instrument selection; expression changes without dynamic changes
	Spatial properties–diverse host environments for each instrument; rhythmic pulses in different stereo locations

THE ARTISTIC ELEMENTS AND THE EXPRESSION OF MUSICAL IDEAS

Throughout history, pitch (in its levels and relationships) has been the most important artistic element of music. The pitch relationships are used to create melodies, harmonies, accompaniment patterns, and tonal systems. Pitch has functioned as the central element in nearly all music that has descended from, or has been significantly influenced by, the European tradition. Pitch is the primary artistic element in much music and is the perceived parameter that contains most of the information that is significant in communicating the message of a piece of music.

Western music might have developed differently. Pitch relationships have been emphasized, much more than other artistic elements, as the primary generator of musical materials, but this did not have to happen. Indeed, the music of other cultures utilize artistic elements in significantly different ways—some emphasizing other elements, many incorporating very different types of pitch relationships.

It is difficult to justify pitch's traditional prominence in expressing musical ideas. While pitch is the perception of the primary attribute of the waveform (frequency, with the other attribute being amplitude), of all the elements it is the most easily detected in many states and values, and pitch is the only artistic element that can be readily perceived as multiples of itself (the octave repetition of pitch levels, and the perceptions of real and tonal transposition of pitch patterns). These attributes, in and of themselves, do not cause pitch to be a more prominent element than the others.

The ability to perceive pitch is not significantly more refined (if at all) than the abilities to perceive the other parameters of sound, especially those parameters that utilize less calculating pitch-related percepts (timbre, environmental characteristics, texture, textural density). It follows that the artistic elements of sound other than pitch, are equally capable of helping to communicate musical ideas. This capability is being realized by the new creative artists who have emerged with audio recording. Creative artists are finding new roles for the artistic elements of sound. The artistic elements that were not available, or that were under utilized, in traditional musical contexts are functioning in new ways in modern music productions. The concept that all of the artistic elements of sound have an equal potential to carry the most significant musical information has been called *equivalence*.

The states of the various components of the artistic elements will make up the musical material. As such, the components of the artistic elements will function in primary or secondary roles of importance in communicating the musical message. It is possible for any artistic element to function in any of the primary and secondary roles of shaping musical materials and of communicating the musical message.

PERSPECTIVE AND FOCUS

The audio professional must be able to understand the artistic elements of sound, how those elements relate to the perceived parameters of sound, and how those two conceptions of sound are used with perspective and focus. The concepts of perspective and focus are central to evaluating sound. The audience will go through this process in a general and intuitive manner; the audio professional must be thorough and systematized when approaching the listening process.

The process allows the listener to perceive the states and activities of the artistic elements of sound as they relate to shaping the musical materials and ultimate message of the work (analytical listening), and as they shape sound quality out of musical contexts (critical listening), or influence the integrity (technical quality) of the sound quality (critical listening).

In order for the message carried by the artistic elements to be perceived, the listener (audience or audio professional) must recognize that important information is being communicated in a certain artistic element. The listener must then decipher the information, to understand the message or recognize the qualities of the sound. The listener will identify the artistic elements that are conveying the important information by scanning the sound material at different perspectives, while focusing his or her attention on the various artistic elements at the various levels of perspective.

The listener is required to identify the appropriate, perceived parameter of sound that will become the center (focus) of his or her attention in deciphering the sound information. Further, the listener is required to determine the appropriate levels of detail that will be deciphered and to extract the information needed to receive the intended musical message.

The *perspective* of the listener determines the level of detail at which the sound material will be perceived. Perspective is the perception of the piece of music (or of sound quality) at a specific level of the structural hierarchy. The content of a hierarchy is entirely dependent upon the nature of the individual work or portion of a musical work.

In a musical context, the detail might break down as follows:

level 1–overall musical texture
level 2–text(lyrics)
level 3–individual musical parts (melody, harmony, etc.)
level 4–individual sound sources (instruments)
level 5–dynamic relationships of sound sources
level 6–composite sound of individual sources (timbre and space)
level 7–pitch, duration, loudness, timbre, space, and durational elements of a particular sound source

level 8—dynamic contour; definition of important components of timbre and space
level 9—definition of prominent harmonics and overtones of the sound source
level 10—dynamic envelopes of prominent overtones and harmonics

Each level of detail represents a unique perspective from which the material can be perceived. Each perspective will allow the listener to observe different characteristics and attributes of the sound material. Perspective might be thought of as a conceptual distance of the listener from the sound material—the nearer the listener to the material, the more detail the listener is able to perceive.

The listener may approach any perspective to extract analytical listening information (pertaining to the function of the musical materials and artistic elements at that level of the structural hierarchy) or to extract critical listening information (pertaining to the definition of the characteristics of the sound itself).

Focus is the act of bringing one's attention to the activity and information taking place at a specific perspective of the structural hierarchy, and/or within a particular artistic element. *Parametric focus* is the concentration of one's attention on a particular component of sound (perceived parameter) within a specific level of the hierarchy. The term parametric focus is used for critical listening evaluations (referring to the perceived parameters of sound, before the context of the music); focus is the term used for analytical listening evaluations.

All parameters of sound need to be scanned, to determine their influence on the integrity of the audio signal; all artistic elements need to be scanned, to determine their importance as carriers and shapers of the musical message. Equal attention must be given to all aspects of sound as, depending on the sound material and purpose of the listening, any perceived parameter of sound or any artistic element may be the correct focus of the listener's attention. An incorrect focus will cause important information to go unperceived and unimportant information to distort the listener's perception of the material.

The listener must develop the listening skills of (1) shifting perspective between various levels of detail, (2) focusing on appropriate elements and parameters at each level of perspective, and (3) shifting between analytical listening (for the importance of the musical material) and critical listening (for the characteristics of the sound itself), to allow the evaluation of sound, discussed later in Part 3, to be accurately performed.

THE SOUND EVENT AND SOUND OBJECT

The artistic elements function at all levels of the hierarchical structure, as they are the elements that comprise the musical materials and sound quality. The concepts of the sound event and the sound object assist the listener in understanding how the musical materials (analytical listening) and sound quality (critical listening) are shaped by the artistic elements at all levels of the structural hierarchy.

These concepts are the perception of an individual musical idea or a sequence of sounds in its entirety. A *sound event* is the shape or design of the musical idea (or abstract sound) as it is experienced over time. The *sound object* is the perception of the whole musical idea (or abstract sound) at an instant, out of time. These concepts are the perception of the form of an individual musical idea. One is conceived as unfolding and evolving through time; the other is conceived as an object that can be conceptualized without the motion of time.

The sound object is the perception of the global qualities of the musical materials, through attention to the artistic elements (or perceived parameters). The sound event is the perception of the musical event (or a sequence of sounds perceived for sound quality), through attention to the artistic elements.

The sound event is a complete musical idea that is perceived by the states and values of the artistic elements of sound, *not* as it is perceived by its component musical materials. The term designates a musical event that is perceived as being extended over a time period and that has significance to the meaning of the work. It is the perception of how a musical idea is created by the artistic elements of sound; it is a perception of how the artistic elements of sound are utilized, to provide the musical section with its unique character.

The term sound object refers to sound material out of its original musical context. For example, in discussing the sound quality of a particular instrument from one recording in comparison to another recording, the instrument would be thought of as a "sound object." A sound object is the consideration of a sound as existing out of time and without relationship to another sound (except a direct comparison of several objects).

The terms sound object and sound event are contrasted at the same hierarchical level (at any level). The concepts of sound objects and sound events allow for analytical listening and critical listening evaluations to be performed, interchangeably, on the same sound materials.

These concepts are able to evaluate the music's use of the artistic elements of sound, in ways that are not necessarily related to the importance or function of the musical materials. Rather, these concepts seek to determine information on the artistic elements (or perceived parameters) themselves, as they exist as single entities (sound objects) and as they change over time (sound events).

THE LISTENER

The audience member and the various audio professionals will have very different levels of listening expertise, and often dramatically different purposes for the listening process.

The *recordist* will have knowledge of the recording process, the states of sound in audio recording, the materials of music, and the hearing mechanism. Further, the recordist will have spent considerable time acquiring the listening skills for evaluating recorded and reproduced music and sound, and musical materials. The recordist is often equally skilled at evaluating the integrity of the perceived parameters of sound and at evaluating the artistic elements of sound and the materials of the musical message.

The *lay listener* will be relying primarily on his or her previous listening experiences. The lay listener may be listening for some meaning in the music and is concerned with the relationship of that musical (or literary) message to his or her personal preferences of musical style and musical and dramatic meanings; the listener may be listening for the sensual aspects of the music; or the listener may be listening for the aesthetic experience. The sophistication of the previous musical experiences of the lay listener is a significant variable that does much to determine the nature of the audience for a particular musical composition.

Great differences exist between individuals from both groups of lay listeners and groups of recordists. The abilities of individuals to understand the significance of musical materials and the artistic elements of sounds, and the physical hearing limitations of certain individuals, cause a generalization of the two groups of listeners to be oversimplified. Instead, the cognitive limitations of the human condition should be explored.

The success of any communication is limited by the receiver. The receiver must be able to accurately process the information in order for communication to occur. Humans are limited in their abilities to understand the content and/or meaning of what they perceive. These limitations are primarily the result of the listener's experience and knowledge, but are also dependent upon the listener's degree of interest in the material, intellect, and physical condition. The same material will yield different information to different listeners (or to the same listener on different listenings), depending on knowledge, experience, analytic reasoning, social-cultural conditioning, expectations of context, attentiveness of listener, and condition of the hearing mechanism.

The listener's accumulated information related to what is being heard, as well as of all subjects related to his or her existence, plays a substantial role in understanding music and sound. *Knowledge* allows the listener to understand a sound, or a musical passage, by relating the experienced sound material to a body of known information. When the listener has a body of known information and/or possible circumstances, the sound event can be matched against those possibilities, to comprehend (and potentially to reason) the meaning of the material.

Knowledge is the amassed body of learned information or known truths from which the listener can draw to make evaluations and judgements on what is being heard. Knowledge areas related to understanding sound (and music) would include acoustics, psychoacoustics, music theory, music history and literature, language, audio recording theory and practice, mathematics, physics, engineering, computer science, communications, and more.

The listener's past *life experiences* are directly related to knowledge. Sound is experienced. In its conceptualized state, sound becomes experienced information. A personal knowledge or experience of the sound is the result of the listening process. Prior listening experiences are a resource that can be drawn from to recognize certain sound events or relationships. Sounds are mapped into memory; the listener is better able to retain sound events in memory when a sound is the same as, or similar to, a sound that has been previously experienced. The act of listening is itself an experience, involving the learning of new information from what is going on in the listener's "present." New information is recognized and understood by comparing it with what has been previously experienced.

The type and quantity of listening experiences, and the personal knowledge gained, will vary significantly between individuals. These listening experiences are significant factors in understanding the messages of music. Different types of music will communicate different messages, and may communicate the message through different musical styles. Difficulties that people experience in understanding or appreciating different types of music can often be attributed to a limited experience with a certain type of music. An individual's listening experiences may have limited their ability to understand the materials (language) of the music, or to appreciate what the music is trying to communicate. Increased knowledge of a type of music, and/or an increased number of experiences in listening to the type of music will increase the listener's ability to understand or appreciate the type of music.

Listening experiences are greatly influenced by the life environment of the individual. The social and cultural environment(s) an individual lives in and has experienced provide opportunities for a certain finite number of listening experiences. Within any environment, certain experiences will occur much more frequently than others. Some types of listening experiences will be very common, some types of listening experiences will never occur.

Social-cultural conditioning will predispose the listener to a certain set of available previous experiences. People are conditioned by their environment (social and cultural), to apply meanings to sounds and to understand stylized musical relationships. They learn to listen for certain relationships in musical materials and the artistic elements of sound. For example, the music of India utilizes pitch and rhythm in significantly different ways than the music of European heritage. Individuals from either culture will not readily understand the meaning or appreciate the subtleties of the music of the other culture, upon initial hearings.

Applying meanings to sounds is the basis for language. Sounds have meaning and can represent ideas. In this manner, a series of short sounds as narrowly defined, isolated ideas can combine (in a prescribed ordering) to create a complex concept. Communicating simple ideas to complex thoughts is thus accomplished by language. It is well known that different cultures have strikingly different languages; some languages have common elements to other languages, and certain languages have elements that are largely unique.

Social context also plays a significant role in defining language sounds and meaning. Quite different meanings may be applied to the same sound, in the same language, by people of the same culture/society. This occurs most often between different social groups (ethnic origins, religious beliefs, age groups, etc.), groups of different economic status, and/or between geographic locations.

Sounds have meanings associated with their source. A sound produced by a car horn will invoke in the listener the thought of an automobile, not the horn itself. Associations between sounds and their sources are largely dependent upon the listener's set of life/listening experiences, as provided by social-cultural environments. One can imagine living conditions under which an individual might never have experienced the sound of a car horn (perhaps an Incan); the sound would not elicit the same response from this person as it would from an urbanite.

The meanings of musical sounds transfer between cultural and social groups in ways very similar to language sounds. Social-cultural conditioning creates expectations as to the function of music. People are conditioned to relate various functions and applications to certain types of music. Dance, celebration, worship, ceremony, accompaniment to visual media, and aesthetic listening are but a few of the functions that music serves in various societies. Each function carries with it certain expectations of musical style; these expectations are defined differently in different cultures and societies.

The life/listening experiences of the listener shape the available resources from which the individual can draw on to understand any sound information. Social-cultural environment conditions the listener by:

1. Providing a predominance of certain listening experiences;
2. Providing certain expectations as to the content of musical materials (the applications of the artistic elements);
3. Providing certain expectations as to the context within which certain types of music will be heard (in church, in a club, in the street, etc.);
4. Providing meanings of association for certain significant sounds (significant sounds being perhaps a siren or a falling tree); and
5. Providing associations of group activity for certain types of music (ceremony, dance, group experiences).

While broadcast media have broadened the number of common elements between social and cultural groups throughout the world, great diversity still exists among human cultures and societal groups. Social-cultural conditioning must remain a significant factor in realizing the limitations of the listener. It might be unrealistic to expect a lay person from the Middle East to understand the musical nuance and message of rock music, just as might be unrealistic to expect an average American, suburban sixteen-year-old to understand the meaning and significance of or appreciate the aesthetic qualities of a Tibetan chant.

Knowledge, experience, and social-cultural conditioning create *expectations* for the listener. The listener will expect to hear certain sounds (or sequences of sounds) under certain circumstances. The listener will expect:

- Certain types of sounds to follow what he or she has already heard (to hear materials in certain relationships);
- To hear certain sounds within a given physical environment (one would not expect to hear a lion sound on a city street);
- Certain sounds in a given musical context (an operatic vocal technique would be unexpected in a reggae work);
- To hear certain musics in certain social-cultural contexts (the listener will expect to hear different music in church, movies, dance clubs, etc.).

When people are presented with something that is not expected, they may be surprised if they are able recognize the material enough to understand it and its context, or may be confused if they cannot recognize the sound or relate the sound to its context. An unexpected sound might intrigue the listener with a unique turn of a musical idea or as a sound slightly out of context (for examples). Conversely, if unexpected sounds that are also little known to the listener are present, the listener will not be able to understand the sound or receive the message of the material, and may likely be dissatisfied or frustrated by the listening experience.

Expected and unexpected sounds and relationships are balanced within all musical styles. A musical style is itself a set of expectations; certain types of musical events and relationships are present that provide consistency and a unique character.

The listener's knowledge, experience, and *analytical reasoning* play important roles in understanding musical messages within various musical styles. Too many unexpected sounds or situations will result in the listener being confused and frustrated. If expectations are filled in predictable ways, the listener will become bored with the material. The listener perceives logic and coherence in the musical materials through fulfilled expectations in the characteristics and functions of the musical materials, coupled with enough unexpected activity to maintain interest.

Listeners use analytical reasoning to extract the meaning of musical materials, when they are unable to identify the material. Analytical reasoning, in music listening, is relating immediate listening experience to knowledge, in a manner capable of deducing meaningful observations and information. The ability of the listener to perform this type of listening activity is dependent upon the listener's intellect, the amount of knowledge that the listener is able to draw from, the listener's previous experience in performing analytic reasoning exercises, and the listener's knowledge of the types of information to extract from the listening experience. This method of listening is similar to analytical listening, as it requires similar skills.

The level of attentiveness of the listener plays an important role in the receiving the musical message. This is exhibited in the states of active and passive listening to music, as well as in the listener's interest in the music.

A difference exists between active and passive listening states. *Passive listening* occurs when the listener is not focused on the listening process or on the music itself in the listening process. Listeners might be listening to music as a background activity (for example, holding a conversation while listening to music), listening to music for its emotive state or feeling, or listening to music for its pulse, or as an accompaniment to dancing. In all of these cases, the listener is not listening to the musical materials themselves, and he or she might not be aware of the music during certain periods of time.

Various levels of detail can be extracted from the *active listening* process. Among many possible states, active listening might take the form of listening to the text and primary melodic lines of a work. It may take the form of following the intricacies of motivic development in a Beethoven string quartet or of evaluating the characteristics of the spectrum of a particular sound object. In all cases, the state of active listening has the listener's attention aware of the states and activities of the musical or sound materials.

The listener is most likely to be an active listener if he or she is interested in the music. Interest in the music may be determined by mood or energy level at the time of the listening experience, but interest is most often associated with the relationship of the music being heard to the listener's preferences (the types of music he or she prefer to hear most often) and the previous experiences that have shaped those preferences.

The final variable between individual listeners is the *condition of the hearing mechanism* itself. Some individuals have impaired hearing. Some people have knowledge of their condition, others do not. The hearing of an individual might vary from the norm because of a defect at birth, from accidental damage because of physical trauma or prolonged exposure to high sound pressure levels, or from natural deterioration caused by the aging process. This variation of the individual's hearing characteristics, from the characteristics of normal human hearing, is of great importance to the recordist. Recordists should have knowledge of the physical characteristics of their own hearing, and make use of this information in evaluating sound in their job functions. Significant hearing problems may make a person unsuitable for many positions in the recording industry.

The *audience* for a piece of music is often considered when predicting the success of the music in communicating its message. The nature of the audience is targeted by defining the knowledge, musical and sociological expectations, and listening experience of the typical audience member. The music conforms to the abilities and expectations of the typical member, to allow the musical message to be successfully communicated and to give the audience a listening experience they might find engaging.

Central to the art of music composition is the intangible balance of providing the listener with enough known material, to allow them to be comfortable in the listening process, as well as enough new material, to allow them to remain interested and engaged in the music. The active listener will seek to discover new detail from the music, while appreciating a known listening situation. In this way, even works that have been listened to repeatedly have the potential to yield more information and allow the listener to remain intrigued by the same piece of music. The relationships of the musical materials, as the music evolves provides the musical experience. The value and success of the perception, and the amount of fascination and personal meaning that the listener finds in the relationships of the musical materials, cannot be calculated or defined; it is another of the intangible aspects of music composition.

TEXT AS SONG LYRICS

When a text is present in a piece of music, it is a significant addition to the musical experience. Through the text, language will be used to communicate a concept or describe a drama within the work, and the sound resources of the language will be exploited, to enhance the aesthetic experience of the music. Songs are often relatively short musical pieces that contain a text; usually a single, rather short text.

The text, or lyrics, of a song is a poem set to music. The text's elements are arranged in some sort of structure (as the structural construction of music), and the concepts of the text will create formal areas that are conceived as a single entity, as well as an overall conception of the text.

The lyrics of songs are constructed in many of the same ways as traditional poetry, which is not intended to be set to music. The primary differences between traditional poetry and poetry as song lyrics lie (1) in the repetitions of certain stanzas or phrases of the poem (unaltered or with slight changes) and (2) in the careful crafting of the meters of the text, the rhythms of the lines, and the timing of the conceptual ideas of the text often found in song lyrics.

The *literary meaning* of the text brings the dimensions of verbal communication of ideas and concepts to the musical experience. Songs have been written on a multitude of subjects–from common, everyday small occurrences to the highest of human ideals. The lyrics of a song might present a story line, be a description of an event, or show the author's feelings about some aspect of the world around them. The text might present the social-political philosophies of the author, or it might be a love song. The potential subjects of a song are, perhaps, limitless.

The presentation of the literary meaning of the text is often enhanced by subordinate phrases or text segments that create new dimensions in the text. These subordinate ideas provide the turns of phrase or concepts that enrich the meaning of the text as a whole. The turning of the phrase allows for different interpretations of the meaning of certain ideas, giving different meanings to different individuals (or groups of people), depending on the experiences of the audience.

The potential for different interpretations allows for some (or much) ambiguity and intrigue in the text. The ambiguity may be clarified with a study of the central concepts of the song lyrics. As is often the case with studying musical materials, reevaluating the concept of a text will often allow the listener to find new relationships of ideas or meanings of materials that enhance the experience of the song for the listener (or recordist).

The concepts used to enhance the literary meaning of the work may or may not be directly related to the central ideas of the text. These ancillary concepts may take many forms and are important in shaping the presentation of the communicative aspects of poetry. A study of poetry or of setting texts to music may be very appropriate for the individual recordist, but it is out of the scope of this writing.

The structure of the text exists on many levels, similar to the hierarchy of musical structure. The conceptual meanings of the text and the sounds and rhythms of the text do not allow for a clear division between the structural aspects of the text and the form-related aspects of the text. The structure of the text should address the sound qualities of the text and its organization of mechanical parts. The form of the text should address the conceptual, often with a recurring concept or theme–a "refrain"–as the song's "chorus."

Some cross-over will occur between the two areas: (1) the structure of the text's presentation may alter the statement's meaning; and (2) concepts can, at times, function as structural subparts. These are the result of the ways people conceptualize in verbal communication, as well as the previous experiences and social-cultural conditioning of the individual.

The components of the structure of a text will be major divisions of the materials of the text and their associated subdivisions (Fig. 3-2). The materials that comprise the components are words, with all their associated meanings and the thoughts and feelings that they invoke from within the individual. Words will be related by their sound qualities, rhyme schemes, rhythms and meters of groups of words, repetitions of words and word sounds, tonal and dynamic vocal inflections, meanings of the words, repetitions of words with different associated meanings, phrases created by the concepts (sentences), and groupings of phrases by subject matter or concepts.

This format will not necessarily be directly transferable to all text settings, but these concepts will provide a meaningful point of departure.

The structures of the text and the music interact in the overall perception of the song; they are perceived as being interrelated, and they serve to enhance each other. The structures may complement one another, or they may serve as areas of contrast, with the text and the music grouped in overlapping segments, unfolding over time.

Both complementary and contrasting relationships of elements of the structures of text and musical materials exist in most works. The two elements play off one another, creating a sense of drama between the text and the music.

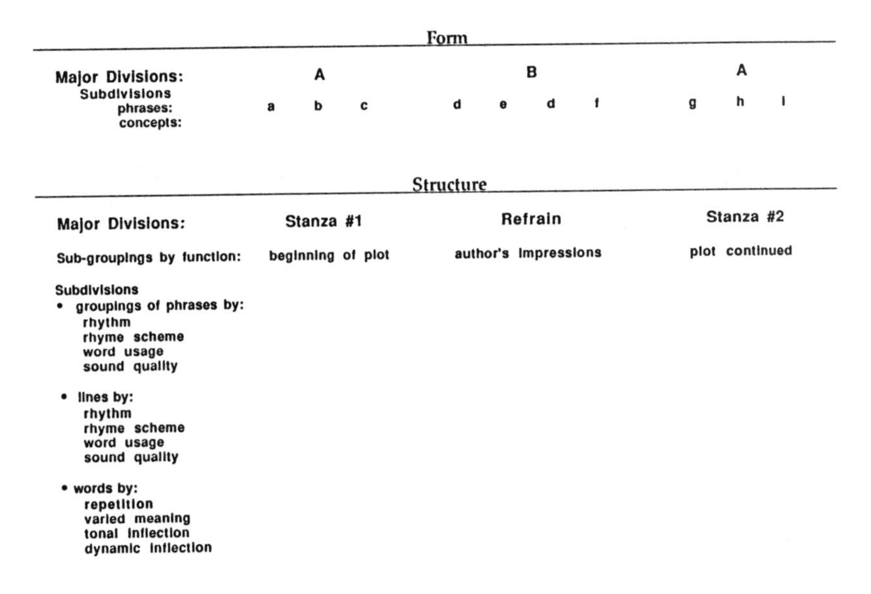

FIGURE 3-2. Form and Structure of Song Lyrics.

The relationships of structures create people's impression of form–their conceptualization of grasping the essence of the entire work in an instant of realization. Within their impression of form as the overall conception of the work, they conceptualize points of climax and points of repose and conceptualize the characteristics of design and shape of the materials that create the movement from one important event, or moment, to the next. People recognize the shape and design of the work as it is represented in their perception of the significant moments of the work, and in the movement between the moments, as they unfold over time.

The relationships of the musical materials create structure in a piece of music; one's perception of the design of structure is his or her conception of form. The structure of a piece of music may be altered significantly, without altering its form. Even when the primary musical materials and the structure of a work are significantly altered, two very different interpretations of the same piece of music will be perceived as being similar when the form (or overall conception) of both performances are similar.

Contrast, for example, two performances of the song "Every Little Thing" one by The Beatles (*Beatles for Sale*, 1964) and by Yes (*Yes*, 1969). The overall shape of the piece is not dramatically altered, but the structures of the two performances are quite different (Fig. 3-3). Great differences exist between the lengths of sections, as well as between the

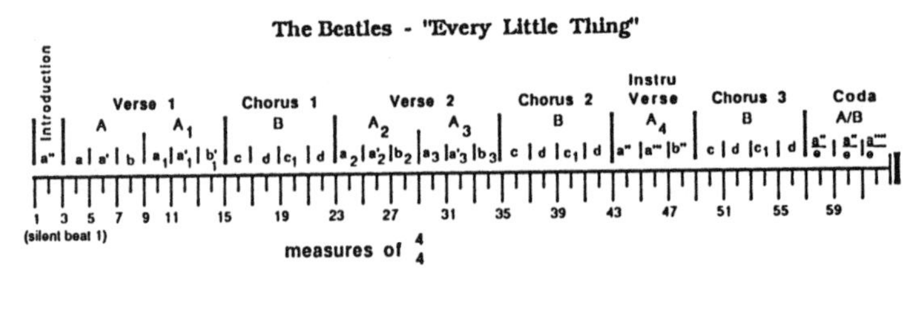

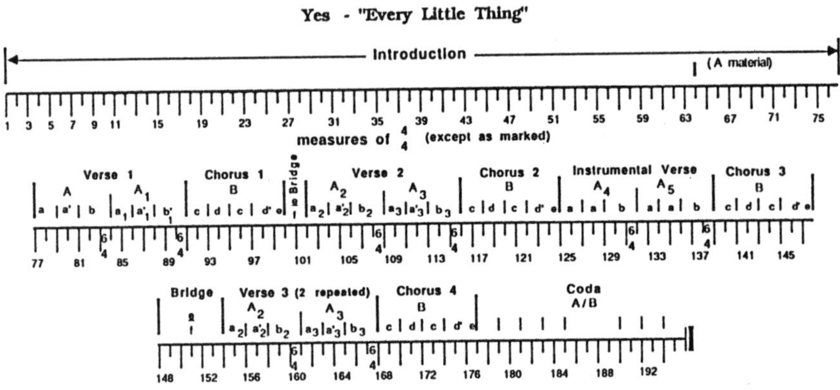

FIGURE 3-3. "Every Little Thing" Performances by The Beatles and by Yes.

treatments of the basic musical materials and how they are organized. Few people would argue that both performances are of the same piece; few people could not perceive dramatic changes in the structure and materials of the two different versions.

II

The Artistic Elements and the Recording Process

4

The Aesthetics of Recording Production

The differences between the various approaches to the aesthetics of recording production can be reduced to two central issues: (1) the relationship of the recording to the live listening experience, and (2) the relationship of the recordist to the creative and artistic decisions of the recording.

The recording process can capture reality, or it can create (through sound relationships) the illusion of a new world. Most recordists find themselves moving about the vast area that separates these two extremes, where they enhance the natural characteristics of sounds. The recordist will determine the appropriate recording aesthetic for the material and function of the current project.

The recordist will use the recording process to support the production aesthetic of the project. The artistic elements of sound will be captured and shaped in relation to the live listening experience and to the musical message. The capabilities of the recording process itself may also be utilized to shape the performance, especially through editing, mixing, micing, processing, and overdubbing techniques.

The recordist will play a more active role in shaping the materials of the music (or the presentation of those materials) in certain projects more than in others. This is often a result of the function of the recordist, in relation to the other people involved in the recording project, in the artistic decision-making process.

For each project individually, the recordist must define their recording aesthetic and their role in the creative process of *The Art of Recording.*

THE ARTISTIC ROLES OF THE RECORDIST

The recordist must have a defined role in the creative process for each project. The project may include the composer of the music, one or many performers, a conductor of the ensemble, and/or a specific recording producer. The recordist must know his or her responsibility to the final artistic product, as well as the roles and responsibilities of the others involved.

The recordist may be functioning to capture the music, as closely dictated by the composer; may be functioning to capture, as realistically as possible, the performance of an ensemble, as precisely directed by the conductor; may be functioning to capture the interactions and individual nuances of a group of performers, without distorting the performance through the recording process; or the recordist may be functioning to provide a recording producer with a precise execution of his or her instructions. In all of these cases, the recordist is allowing the artistic decisions of others to be represented as accurately as possible in the recording. The recordist's role is to facilitate the artistic ideas of others, and not to directly impose his or her ideas onto the project.

The recordist might offer suggestions to the creative artists, or take an active role in the artistic decision-making processes. The recordist might be active in shaping a performance of an existing work, or in creating a new piece of music. The recordist might be active in determining the sound qualities of the instruments of the recording, or in determining the sound sources themselves. Vastly different levels of participation in the artistic process are required, from one project to the next.

Writing a piece of music for a recording is often a collaborative effort. It may be between many people (composer, performers, producer, recordist) or just a few (performer/composer and recordist/composer).

In many ways, the recordist is a creative artist. The recordist may serve the traditional functions of a composer, a conductor, or a performer. The recordist may also shape sounds in new ways; he or she has new controls over sound and live performances that allow for a new musical voice. It is possible to compose with the instruments of the recording studio, to shape sounds or performances through the use of recording techniques, or to create a new musical environment for someone else's musical ideas and performances.

As will become clear later, the recording studio can be thought of as a musical instrument or a collection of musical instruments. In this way, the recordist may "conduct" all the available sound sources (for example, bringing sounds into and out of the musical texture through mixing), "perform" the musical ideas through the recording process, alter or reshape the sounds of the sources or "interpret" the musical ideas in ways that are not possible acoustically, or create ("compose") new musical ideas or sounds.

THE RECORDING AND REALITY: SHAPING THE RECORDING AESTHETIC

The role of the recordist may be to capture a live event as accurately as possible, in relation to the dimensions of that real-life experience; to alter the artistic elements of sound, to "enhance" the quality of that real-life experience; or to create a new "reality" or set of conditions for the existence and relationships of sounds. Reality is simulated, enhanced, or "created" through the recording process.

The relationship of the recording to the live listening experience is central to the aesthetic quality of the recording. A recording may differ from the live listening experience by (1) utilizing the artistic elements of sound in ways that cannot happen in nature, and (2) presenting impossible human performances and compilations of "perfect" performances.

The artistic elements that most influence the lifelike qualities of the recording are environmental characteristics, the dimensions of the sound stage, and the relationships of musical balance to the timbres of sound sources.

Sound exists in space. Humans conceptualize sound, especially in the context of music performances, in relation to the spaces in which the sound exists. The recording process must provide the illusion of space, to convince listeners that the sound has been reproduced in a way that is associated with their reality. The recording will provide the illusion of a performance space or a physical environment for the performance.

The recording will provide the illusion of a performance space, wherein the recording can be imagined as existing during its re-performance (playback). The realistic nature of the performance of the recording will play a central role in establishing the relationship of the recording to the live listening experience.

The listener will imagine the recording as a performance in some invisible space. Since humans evaluate sounds in relation to the environments within which they are heard as existing, the listener will subconsciously scan the recording to establish environmental characteristics, an imaginary stage (sound stage), and a perceived performance environment. This information allows the listener to complete the process of establishing a "reality" for the listening experience of the recorded music performance.

The imaginary environments will be the captured reality of the original performance space, an altered or enhanced reality of the original performance space, or new realities that are created for the performance through signal processing. The listener will make fundamental judgements about the material of the recording, based on the qualities of the environment(s) simulated in the recording, and will match the musical material against the appropriateness of the environmental cues.

The listener will imagine the location of the sound sources relative to one another and to the overall environment. He or she will envision the sound stage of the recording. In so doing, the entire ensemble will be placed at a certain distance from the listener, and each individual sound source will be placed at a distance and at an angle from their perceived location. The relationship of these cues to the potentials of live performances will also define the aesthetic of the recording, in relation to the possibilities of reality.

The sound stage of the recording might place the sound sources in locations that purposefully resemble those of a live performance. In certain recording techniques, the integrity of this imaging is a primary concern. Certain stereo microphone techniques are designed to accurately capture the depth of the sound stage and the lateral location of the sound sources; other techniques accurately capture the microphone-to-stage distance and stage width.

The sound stage is often altered by the recordist, to enhance the musical material. One or several additional microphones may be used to accent certain members of an ensemble. The highlighted instruments are given a distance from the listener, a width of image, or a specific location that provides them with more prominence in the musical texture. This can be performed subtly, so as not to dramatically alter the natural qualities of the recording, or it can be quite pronounced, depending on the aesthetic of the recordist.

Sound stages are created for multitrack recordings and for electronic music recordings. These recordings were created outside a common environment, with minimal naturally occurring spatial cues captured with the sound sources. The recording will be given spatial cues by either the recordist (through the recording process) or the listener's imagination. The recordist will create his or her own sound stage for the recording.

The recordist can provide the sound sources with lifelike environments, placing them in natural physical relationships to one another, or can purposefully create environmental, distance, and localization cues that would be impossible in nature. In the latter, the recordist is most commonly interested in assigning environmental qualities that enhance the sound quality of the sound source as a primary consideration; the actual spatial illusions currently are often of secondary importance. The degrees to which the sounds have unnatural spatial relationships may be very subtle or very pronounced.

The interrelationships of musical balance and the perceived differences of sound quality of sound sources played at different dynamic levels (performance intensity) are integral parts of live performances and are easily altered by the recording process.

Recordings that attempt to capture the aesthetics of the live performance will seek to capture the musical balance of the performers as they (or the conductor) intended. The changes in the sound quality of the instruments will be precisely aligned with changes of dynamic levels in the musical balance of the ensemble and with changes in musical expression.

Recordings that seek to enhance the characteristics of the live performance may contain slight changes in musical balance that were not the result of the performers, but of the mixing process. These alterations will be heard as changes in dynamic levels that are not supported by corresponding changes in the sound qualities of the instrument(s). This enhancement might take place in only a few instruments, or it may be used extensively throughout the entire ensemble. This enhancement technique may be quite subtle and difficult to detect, or it may be prominent.

Alterations in dynamic levels, and thus musical balance, that are not aligned with perceived changes in performance intensities have become integral parts of music written for recordings. Multitrack mixes frequently exhibit changes in musical balance that were not caused by the performers. These changes in dynamic levels are inconsistent with the sound qualities of the instruments in the final recording. This inconsistency of one element tracking another enhances the potential of each element to be used individually in shaping or enhancing the musical material.

The relationships of musical balance to the timbre of sound sources in many multitrack recordings creates a wealth of contradictions between reality and what is heard. The aesthetics of this type of recording leans towards redefining reality, with each new project, and is a stark contrast to the aesthetic of trying to capture the reality of the live performance.

THE RECORDING AESTHETIC IN RELATION TO THE PERFORMANCE EVENT

The recording medium is often called upon to be transparent. In these contexts, it is the function of the recording to capture the sound as accurately as possible, to capture the live performance without distortion. This type of aesthetic is common for archival recordings that document events; the recording seeks to accurately capture the particular music performance. The recording may or may not be sensitive to the performance environment. Often, these recordings attempt to capture the sound of the music performance, without considering the artistic dimension of the relationship of the music (and musicians) and the performance space (and audience), or they seek to negate any influence of the performance space on the sound of the recording.

Because these are recordings of live performances, the recordist is not involved with compiling the performance. The performance takes place in real time; it will not be possible to back up and fix a certain section or idea. The recordist is primarily concerned with the technical aspects of the sound of the recording (critical listening) and the sound quality of the overall ensemble (at the highest level of perspective).

A limited number of microphones are used in making this type of recording. Usually, two microphones are utilized in some appropriate stereo microphone technique, placed close to the ensemble. The microphones generally are sent directly to a two-track mastering recorder, with little or no signal processing. The recordist has little immediate control over the quality of the sound and the shaping of the performance.

The recording medium can be transparent in documenting a performance, while placing the music in a complementary relationship with the host environment of the performance. Specific pieces of music are best suited to certain environments and are most accurately perceived from certain listening distances. The artistic message of a piece of music will be most effectively communicated in a specific environment, with the listener at an ideal distance from the ensemble.

Recordings can ensure that pieces of music will be perceived as having been performed in an "ideal" environment, with the listener located at an "ideal" distance from the ensemble, when listening to the recording. This locating of the listener at the *ideal seat*, can be accomplished without altering the performance itself, while keeping the recording process transparent.

The recordist will determine the type and amount of influence that the acoustic performance environment will have on the final recording. Microphone selection, choice of stereo microphone array, and array placement within the performance environment are the primary determinants of the environment "sound" that is captured from the performance environment. The characteristics of the environment may be sensitively enhanced by artificial reverberation units or other time processors. The distance of the listener from the ensemble is determined primarily through microphone placement and time processing.

This recording aesthetic attempts to present the music in the most suitable setting possible for that particular work, and to simulate the listening experience in the concert hall. The recordist seeks to ensure that the sounds will be in the same spatial relationships as the live performance, the balance of the musical parts will not be altered by the recording process, and the quality of each sound source will be captured in a consistent manner.

This aesthetic can have the recordist more involved with the decision-making process in some projects than in others, and may be used for live concert recording, as well as session recording. This aesthetic may be used for a wide variety of music. While it is common in orchestral and other art music formats, it is equally appropriate for jazz or any other music recordings. The performers must be refined in their sensitivity to the other performers, their control of their musical materials, and in relationships to the musical materials to the whole ensemble. In session recordings, some (or much) editing may be a part of this aesthetic; a consistency of sound quality and spatial relationships between all portions of the work will nearly always be sought.

The recording medium may be utilized to enhance the performance in widely varying degrees. This aesthetic may be a slight extension of the concept of a transparent live recording, with the recording process slightly enhancing certain musical ideas, or it may set another extreme of being a lifelike session recording that was recorded out of real time.

This aesthetic simulates a natural listening experience, by capturing or creating many of the experience's inherent characteristics. The timbre and dynamic relationships, spatial cues, and editing techniques all serve to create the impression that the recording could take place within reality.

When this aesthetic is an extension of the concept of a transparent live recording, sounds are placed in the sound stage of the recording in the same relative positions as the instruments were in during the recording. The width and depth of the sound stage are realistic, and the recording will usually have a single environmental characteristic applied to the overall program (a single soloist might be present with a different environmental characteristic). Dynamic changes are nearly always aligned with timbre changes (some microphone highlighting might create a limited number of dynamic changes without timbre changes). The recording process is used to slightly enhance the musical ideas of the live performance.

This aesthetic may be used for controlled live performances (those that have been rehearsed with the recordist) or in recording sessions, for a wide variety of musical styles. Minimal micing will be used, usually an overall stereo array with a small number of accent microphones (or stereo pairs). Accent microphones allow this aesthetic to be adaptable to stage recordings of musical theater and opera. Recording is usually mixed directly to a two-track mastering deck, with mixing decisions taking place during the rehearsals or during recording sessions. Recording submixes to a multitrack recorder is also common, but many of the same decisions related to the sound of the recording as above are still accomplished during the recording session or rehearsals.

Recording sessions will often be comprised of many takes of large and small sections of the work. As the ensemble balance is largely controlled by the performers and the parts are not singled out (making rerecording of individual parts unavailable), ensemble problems of accuracy and sound quality often cause a lengthy recording session and a large set of session master tapes.

The master tape will be a collection of the many takes. Editing these many takes of the musical material becomes an integral part of the recording process. The recordist's goal is to compile a *perfect performance* of the work.

The aesthetic of slightly enhancing the reality of the performance may also be found in session recordings that simulate natural sound relationships. Although recorded out of real time, the recordings will simulate the experience of live music. Some emphasis of certain musical materials (and/or artistic elements) over others will be unavoidable in the recording process and will diminish the naturalness of the relationships of the sounds. Some recordings may simulate reality only generally, but their inherent conception is still to provide the illusion of a naturally occurring performance. Even with the complete control of the multitrack recording (micing, processing, and mixing) process, the goal is to provide the illusion of a live performance. In all other ways, music written for recording might follow the most unnatural recording processes and conceptions.

The music written for the recording medium may be significantly different from traditional music. It may be constructed in new ways and contain new artistic elements. Music written to be recorded, and especially music written during or through the recording process, is often composed and/or performed in layers.

The musical materials are often written and recorded one part at a time, or a small group of instruments at a time. The recordings use close micing techniques that ensure a separation of parts (and thus allow for precise control of the individual sound source) or that will physically isolate the performers/performances from one another. The parts are continuously compiled on a multitrack recorder, with each new musical line added to the musical texture. Players often perform their parts many times; any number of versions may be recorded before the desired result is achieved. The recordist may be responsible for listening for performance mistakes, listening for the most interesting and successful performance, keeping track of which portion of which musical part was performed most accurately, on which take, and so on.

The final piece may be a composite of any number of performances, or it may be a controlled integration of many different musical ideas and personalities. The performances may or may not have taken place at the same studio or during the same day (or year); the performers may or may not have met and discussed their musical intentions.

The recording medium can create the illusion of a performance that contains characteristics that cannot exist in nature. This aesthetic has become common since the mid-1960s. In this new aesthetic, the recording medium is utilized for its own creative potentials. The recording medium is a new musical ensemble or a new set of resources, for shaping a performance or creating a musical composition.

Music written to be recorded may exploit environmental characteristics, musical balance and sound quality contradictions, sound stage depth and width, sound source imaging, or other unique elements to create, define, or enhance its musical materials.

This aesthetic might purposefully create relationships that cannot exist in nature: a whisper of a vocalist might be significantly louder than a cymbal crash. This aesthetic will use the unique qualities of recorded sound in communicating the work's musical message.

Recordings of this aesthetic seek to create a new reality for each work or project. New relationships of sound are calculated and incorporated into the music. Recordists (engineers and producers) develop personal styles of the ways that they shape aspects of balance, imaging, sound stage, and environment, while continuing to explore the expressive potentials of recording and the medium's relationships with reality.

Much of today's popular music falls within either the above aesthetic or the aesthetic of using the recording medium to enhance the illusion of a live performance.

Most of the works discussed in Part 3 will be of one of these two aesthetics, as the artistic elements are most easily observed in these types of music recordings. Many of the artistic considerations of the recording process are directly contained in these two aesthetics.

ALTERED REALITIES OF MUSIC PERFORMANCE

The listener's perception of the reality of the music performance event itself is also altered by the aesthetics of recording production. The recording provides an illusion of a live performance; the content or qualities of the perceived music performance may vary, from a slight improvement of our listening realities to being a live performance that is existing in ways that are impossible in our known world.

Recording allows a music performance to be an object that can be precisely polished by the artists, physically held in one's hands, and owned by a member of the general public. The reality of a live music performance as an experience witnessed in a fleeting instant in time, and as retained only in the memories of those who experienced the event, is significantly altered by the recording process. The recording is a *permanent performance* of the piece of music.

New pressures, ideals, and aesthetics are placed on the artists responsible for the individual recording. A recording may distort the live listening experience by:

1. Presentating humanly impossible performances;
2. Providing performance conditions that are inconsistent with reality;
3. Presentating error-free and precisely crafted performances; and
4. Providing a permanent record of a music performance.

A recording is a permanent performance of a piece of music. It is a period of time that has been created or captured and that may be preserved forever. The performance can be revisited (and observed at any level of detail) at any time and any number of times.

Recordings are often conceived as being *definitive performances* of pieces of music. The definitive performance may be conceived as being either that of a certain artist or of the particular piece of music. An artist's performance/recording of a work might be what is widely accepted as the definitive performance (or reference) of how the work exists in its most suitable form, in relation to performance technique or to the communication of the musical message. A specific recording of a work can also serve as a definitive reference of how a work exists in its most suitable state and in relation to recording practice or to musical considerations.

Not only are recordings a means of creating an art form, they also preserve the artistic ideas of music performance and expression that do not rely on the recording process. Recordings may permanently preserve the music performances of any artist. Recordings may provide historical documentation or archival functions by preserving the music performances of particular artists, ensembles, events, and so forth.

The great contradiction of producing a recording that is a permanent record is that the recording often becomes dated. Artists grow; their musical abilities, levels of understanding, artistic sensibilities, and musical ideas change. The permanent performance that was previously created (perhaps only a few weeks before) may no longer be representative of the artist's abilities or aesthetic opinions.

The recording will often represent the artists' and recordist's conception of a perfect performance of the work. Theoretically, a perfect performance of any piece of music can be produced through the recording process. The definition of the perfect performance may vary considerably between performers, but the concept of the recording itself will be similar: a presentation of what the performers and producers of the recording perceive as being the most appropriate interpretation of the piece of music, under the most appropriate performance conditions (instruments used, performance space, etc.).

A perfect performance will combine the artists' desired interpretation of the music (and an absence of performance inaccuracies) with an illusion of the drama of a live concert, as being experienced at the "ideal" listening location of an "ideal" performance environment for the ensemble and piece of music. Often, practical considerations of the recording process will compromise the actual quality of the recording, but the goal of the recording remains constant.

The recording may present musical ideas and sound qualities and relationships that are impossible to create in live performance. Musical materials may be presented in ways that are beyond the potentials of human execution: rapid passages performed precisely and flawlessly, dynamics and sound quality expressions that change levels quickly and in contradictory ways, and using a single human voice to perform many different parts are but a few of the possibilities. Humanly impossible performance techniques and relationships are easily created through utilizing the recording processes.

The reality of what is humanly possible in music performance is often inconsistent with the music performance of the recording. The relationships and characteristics of sound sources, the interrelationships of musical ideas and artistic elements, and the perceived physical performance of the musical ideas may be such that they could not take place in nature. It could not be accomplished without recording techniques and technologies.

Recordings have greatly influenced people's expectations of live music performance. The listening audience of a recording becomes accustomed to a recording as being the "perfect performance" of a piece of music. The audience may learn the subtleties of a recording quite intimately.

A music recording is actually a particular performance of a piece of music that has been created or captured. When an audience member owns a copy of the recording, or when a recording has received much exposure through media, the audience may have listened to a recording many, many times. The recording becomes the definitive performance of the music, for some audience members. The audience will carry this knowledge of the music recording into a live music concert, and may impose unrealistic expectations onto the performers and the event.

The artists' new and different interpretations of the music, the absence of recording production techniques, and the inconsistencies of human performance will create differences between the live performance and the known recording of a piece of music. A potential exists for audiences to become less involved with the drama and excitement of the live performance. Audiences may attend concerts to hear, publicly and with live performers, the music performances (recordings) they have come to know well as recordings. Audience members do not always accept the reality that the live performance was not the same as the studio-produced recording; a potential exists for the audience member to be dissatisfied that the definitive performance that they know well was not reproduced for them at the live event.

Unrealistic expectations may be placed on the performers by an audience; they may be expected to perform flawlessly, or with the same version and interpretation of the work as their released recording. An audience may expect the performing artist to reproduce one particular performance of the work. The performer may be restricted from allowing their interpretation of the music to evolve and change according to their growing experience, and may be restricted from creating a more exciting performance, by reacting to the audience. The subtleties of artistic expression that are possible only through the artist and audience interaction, along with other unique qualities of live music performance, may be lost from an event and diminish the musical experience.

It is unrealistic to expect to hear a precise reenactment of a recording in a live concert environment. Many music recordings have been produced in such a way that a live performance of all musical parts, sounds, and relationships is impossible. These are potential negative outgrowths of the audience's familiarity with certain recordings and the new listening habits afforded by readily available music performances.

The general public hears much more recorded music than live music. They are prone to incorrectly use the same criteria in appreciating both the live and the recorded performance. People own their own personal copies of performances. A tendency to personalize or to become attached to those performances is common—especially among younger, less experienced listeners. When the performance is changed, something personal ("their" music) has been altered.

The recording aesthetic is determined by the relationship of the recording to the live listening experience. The recording aesthetic is arrived at through a careful consideration of the musical material, the function of the recording (sound track, advertisement), and the desired final character of the recording. The recordist's role is defined by his or her contributions (or lack thereof) to the process of making the creative decisions of the recording.

The recordist's controls over the many qualities of the final, music recording are highly variable. The recordist is responsible for the overall characteristics of the recording and may be in control of (and responsible for) its most minute details, depending on the recording techniques being utilized. The recordist might have precise control over shaping or creating a performance, or may be merely capturing the global aspects of a live performance.

The amount and types of control that the recordist utilizes in the recording process will determine the degree of influence he or she has on the final content of the music recording. The recording medium may be used to greatly influence the sounds being captured by the microphones, or it may shape sound much more subtly. The recording process will be used differently, depending on the particular project.

The aesthetics of recording production vary with the individual, with the musical material, and with the artistic message and objectives of a certain project. An aesthetic position or approach may or may not be appropriate for a certain context. It may enhance the artists conception of the music, or it may not. It may be consistent with other considerations of the project or the music, or it may not (perhaps consistency is not desired).

The aesthetic approach to recording production creates a conceptual context of the artistic aspects of recording. The intangible aspects of the art can then be appreciated within this context.

5

Preliminary Stages: Defining the Materials of the Project

The recordist must learn to anticipate the content and scope of the recording project. Before sounds are dedicated to tape (or disc), the recordist must have some clear ideas of the dimensions of the project. The recordist will then be able to correctly make the choices of the preliminary stages of a music production or audio project.

These choices will affect the entire project. If they are made poorly, the entire project may be compromised. The recordist will be able to anticipate the problems of each stage of the recording project, by considering the state and sound qualities of the project at all the various stages of the sequence of creating a music recording.

In the preliminary stages, the materials of the project are defined. These materials comprise the creative resources of the work. The recordist will often be responsible for making many of these decisions, depending on their role in the decision-making process of the artistic aspects of the project. The choices that the recordist may be required to make during the preliminary stages of the project are choice of sound sources, choice of how to capture the sound sources (initial shaping of the sound sources), and choice of how the sound sources will be heard during the recording process (monitoring format).

The tracking process must often be planned during the preliminary stages of a recording project. This topic is covered in the next chapter. The decisions of tracking will, however, often occur during the preliminary stages, before sounds are dedicated to tape. These include:

- Defining the number of available tracks;
- Track assignments of the material on the multitrack tape;
- Mixing microphones during the tracking process;
- Placing sync tones;
- Performance order planning;

- Planning reference tracks; and
- Rehearsing the session.

SOUND SOURCES: ARTISTIC RESOURCES AND THE CHOICE OF TIMBRES

The music to be recorded and a certain set of sound sources (voices and instruments) to perform the music are usually dictated to the recordist as dimensions of the project. Decisions on selecting sound sources and on many of the supportive aspects of the music are, however, often made during the preliminary stages of the recording process (or during the production process itself).

The selection of sound sources will define the materials of the music production. The sound sources will be the vehicle that presents the musical ideas. They must be selected carefully, with attention to their anticipated roles in the production. Sound sources are often coupled with the musical ideas themselves, as the primary artistic resources of the music.

In selecting sound sources, decisions are being made as to the most appropriate timbre, to present the musical materials. In doing so, the recordist will consider the sound source in relation to:

- Suitability of its sound quality to the musical ideas;
- Potential to deliver the required creative expression; and
- Pitch area information for anticipated placement in relation to textural density.

Individual performers may be selected because of their unique sound qualities. Individual performers, themselves, are unique sound sources. This is especially true of vocalists, who are sought for their unique singing voice and styles, as well as for their speaking voices.

Accomplished instrumental performers that have developed their own style(s) of playing or that have advanced performance technique are also sought for their uniqueness as sound sources. Individual performers often bring their own creative ideas and special performance talents to a project, and give considerable aid in defining the sound qualities of the sound sources.

The selection of particular performers for a recording is important in defining the sound quality of the sound source, down to the most minute detail. The recording is a permanent performance of the piece of music. As such, the selection of performers for a performance is often an important consideration in selecting the sound qualities of the sound sources.

Just as live, human performers function as unique sound sources in a music recording, nonhuman performers are able to function in the same ways. Computers can be and sequencers are nonhuman, mechanized performers.

Computers and sequencers have the potential to be programmed in great detail. They are capable of performing complex musical materials and ideas by controlling (sending performance instructions to) sound sampling and synthesis devices. Computers and sequencers (both hardware and software) have the potential to give certain characteristic sound qualities to the music, but may also be very lifelike. They are capable of very detailed and precise control over a performance (and of making related timbre changes), often providing very human-like results.

Sound sources and sound qualities can be created. The sound manipulation and generation techniques of sound synthesis allow for the design of sound sources. Many approaches to sound synthesis are available:

- Analog synthesis techniques;
- Additive and FM digital synthesis techniques;
- Many hybrid (analog + digital + sampling) synthesis techniques (such as waveshaping, phase distortion, wavetable, physical modeling, granular synthesis techniques, etc.);
- *Musique concrète* techniques;
- Recording and performing techniques on sound samplers (sampled live or with commercially available sound libraries); and
- Computer generated (often employing typical digital synthesis or sampling techniques).

Creating sound sources allows the recordist great freedom in shaping sound qualities. The recordist will be functioning as a *sound designer*, whose goal is to create a sound (with a sound quality) that will most effectively present the musical materials and the ideas of the music. Sound sources that precisely suit the contexts of the sound and the meaning of the music may be crafted or created by the recordist.

While an examination of the sound synthesis process is out of the scope of this writing, it is important for the recordist to be aware of the many creative options afforded by sound synthesis. The study of sound synthesis from the perspective of building timbres will greatly assist the recordist in understanding the components of sound and how these components may be used as artistic elements. Signal processing and sound synthesis share many common traits.

By creating sound sources, the recordist will be presenting the audience with unfamiliar "instruments." The sources (new instruments) may be performing significant musical material. The reality of the performance has been altered out of the direct experience of the listener. The recordist must be more aware that he or she will be creating a new reality of sound relationships or might need to emphasize known sound relationships in order to reestablish known experiences. These relationships will be accomplished in such a way as to support the musical materials and ideas of the recording.

The human realities of sound relationships are most closely associated with acoustical environments. The listener will process the characteristics of the environment within which the sound source is sounding, as well as the location of the source within its acoustical environments, to imagine the reality of the performance. The acoustical environment itself will also function as a sound source, of sorts.

A set of environmental characteristics may be so much a part of the sound quality of a sound source that the actual components of timbre become secondary in the global impression of sound quality judgements. The acoustical environment, in essence, becomes the sound source that is projected by the instrument that it contains. This is an unusual sound occurrence that is accomplished through a very high percentage of reverberant sound over direct sound, often causing the sound to appear to be "otherworldly."

Nonmusical concepts often find a place in a music project. As sound sources, speech and special effects require special consideration.

With speech as a sound source, a particular voice is selected to complement the meaning of the text and to complement the other sounds in the musical texture. The voice is carefully selected for the appropriateness of its sound quality, and thus its dramatic or theatrical impact on the meaning of the text to be recited.

Special effects are sound sources that are used to elicit associated thoughts from the listener. The associated thoughts generated by the special effects are *not* directly related to the context of the particular piece of music. Special effects pull the listener out of the context of the piece of music, to perceive external concepts or ideas. A horn sound occurring in a piece of music, used to elicit the mental image of an automobile, is an effect; the same sound used as part of the musical material, to complement the musical ideas of the work, would *not* be a special effect, but rather a musical sound source.

Sounds for special effects and for sampling/synthesis devices may be obtained from sound libraries. Sound libraries are collections of sounds available on tape, compact disc, or lp, or in a proprietary digital format (for specific sampling-related devices). They are a resource to provide a wide variety of sounds from virtually every conceivable source (natural and created sounds), and many different types of presentation (light through heavy footsteps of the same person, a child's footsteps, an elderly person's footsteps, high-heeled shoes, etc.). Individual sound libraries are commercially available and allow the recordist to purchase specific types of sounds for specific projects.

MICROPHONES: THE AESTHETIC DECISIONS OF CAPTURING TIMBRES

The interactions created by selecting a particular microphone, placing the microphone at a particular location (within the particular environmental conditions), and matching those decisions to the characteristics of the particular sound source are major determinants of the sound quality of sound sources in music production.

A specific microphone will be used to make a certain recording of a sound source because of how its performance characteristics complement the sound characteristics of the sound source. This interaction of the characteristics of the microphone and the sound source allows the recordist to obtain the desired sound quality of the recorded sound source.

A microphone will be selected for a particular recording because it is the most appropriate for the desired, final sound quality. The recordist will determine which microphone is the most appropriate by comparing the contributions of the sound quality of the sound source and the performance characteristics of the microphone to the recordist's conception of the final sound quality to be recorded. The recordist will also evaluate the selected microphone against the practical limitations of placing the microphone in the recording environment, and in relation to the sound source to be recorded and sound source(s) *not* to be recorded.

No single microphone will be the "best" microphone for every sound source, or for the same source for every piece of music. The microphones selected for recording the same sound source will vary widely, depending on the above circumstances, on the desired sound, and on what microphones are available.

All microphones can be evaluated by their *performance characteristics*. These characteristics give information on how the microphone will consistently respond to sound. Thus, through these characteristics, the recordist can anticipate how the microphone will transform the sound quality of the sound source, while it is being recorded.

As the microphone alters the sound source, it has the potential to contribute positively in shaping the artistic elements. If the recordist is in control of the process of selecting the appropriate microphone for the sound source and conditions of the recording, the selection and applications of microphones can be a resource for artistic expression. The artistic elements of the recording can be captured (recorded) in the desired form, and the microphone and its placement will become part of the artistic decision-making process.

A number of microphone performance characteristics are most prominent in shaping the sound quality. These characteristics are of central concern in determining the artistic results of selecting these microphones for certain sound sources. These microphone performance characteristics are:

- Frequency response;
- Directional response;
- Transient response; and
- Distance sensitivity.

Frequency response is a measure of how the microphone responds to frequency levels. Amplitude differences at various frequency ranges are determined throughout the audio range, to calculate the sensitivity of the microphone to frequency.

The frequency response of a microphone is often comprised of certain frequency bands that the microphone will accentuate or attenuate. Matching a sound source with similar frequency characteristics may or may not provide the recordist with the desired sound. The microphone may accentuate certain characteristics of the sound source; perhaps a microphone with somewhat opposite frequency characteristics as the sound source will be a more appropriate choice. Again, this decision is dependent upon the final, desired sound.

Microphones do not respond equally to sounds arriving at different angles to its diaphragm. The *directional response* of a microphone is its sensitivity to sounds arriving at various angles to the diaphragm. The *polar pattern* of a microphone depicts the sensitivity of a microphone to sounds at various frequencies in front, in back, and to the sides; the actual pattern is spherical around the microphone (Fig. 5-1). Directional response measures the microphone's sensitivity to sounds arriving from angles, but calculates this sensitivity at only a few frequencies.

Sounds directly in front of the microphone diaphragm are considered to be *on-axis*. Sounds deviating from this zero-degree point on the *polar curve* are considered *off-axis* and are plotted in relation to the on-axis reference level. The frequency response of most microphones will vary markedly to sounds at different angles. Even microphones that show no pronounced frequency areas of accentuation or attenuation (flat frequency response) on-axis, will show an altered frequency response at the sides and the back of the polar pattern.

These variations in frequency response at different angles are commonly

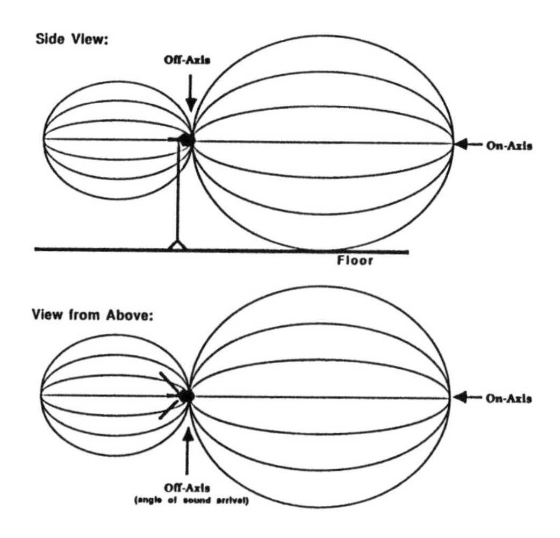

FIGURE 5-1. Polar Pattern Spheres and the Microphone Axis.

called *off-axis coloration.* These changes are more pronounced at or towards the attenuated angles of the microphone patterns. In the intermediate angles between directly on-axis and the dead areas of directional patterns, a slight change in the angle/direction of the microphone can make a substantial difference in the frequency response of the captured sound source. Frequencies of about 4 kHz are most dramatically altered by slight angle changes.

The amount of off-axis coloration is an important measure of the microphone's suitability to a variety of situations. This is especially pronounced in stereo microphone array recording techniques. Instruments at the edges of the array's pick-up pattern and the reverberant sound of the hall will arrive at the array mostly from angles that are off-axis. The sound qualities of those instruments and of the reverberant energy may be altered significantly by the off-axis coloration. Off-axis coloration may have a profound impact on the sound qualities of the recording.

Microphones have different response times, before they begin to accurately track the waveform of the sound source. This *transient response* time distorts the initial, transient portions of the sound's timbre. Slow transient response is most noticeable when the microphone is applied to a sound source that has a fast initial attack time, in the dynamic envelope, and a large amount of spectral energy during the onset.

Transient response is a microphone performance characteristic that is not included in the manufacturer's specifications information that accompanies promotional literature and owner's manuals, and is not governed by a standard of measurement.

The acute hearing of the recordist will be the primary judge of this microphone characteristic. The recordist must become aware of differences in timbre that are present in the early time field of the miked sound source, as compared with the live sound source. These differences are derived through the critical listening process and are vitally important to the recordist, who must be in control of capturing and shaping the sound as desired. Two microphones with identical frequency response curves may have completely different sound characteristics caused by different transient response times.

All microphones will respond differently to the same sound source, at the same distance and angle. The ability of each microphone to capture the detail of a source's timbre will be different. Microphones will have different sensitivities in relation to distance. The *distance sensitivity* of a microphone is often influenced by the polar response of the microphone and/or its transduction principle (condenser, moving coil, ribbon, etc).

Directional patterns are often able to capture timbral detail of a sound source at a greater distance than an omnidirectional microphone. This is primarily the result of the ratio of direct to indirect sound and a masking of timbre detail, but it may also be attributed to transduction principle, depending on the particular circumstances. Similarly, condenser microphones will have a tendency towards greater distance sensitivity than dynamic microphones, due to its more sensitive transfer of energy. Many times a microphone with a small diaphragm will have greater distance sensitivity than a microphone with a larger diaphragm, all other factors being equal.

The concept of the distance sensitivity (sometimes called "reach") of a microphone is an important one. The recordist must judge this microphone characteristic through acute listening and experience. The recordist must become aware of differences in timbre detail that are present between the miked sound source and the live sound source. Distance sensitivity is a microphone performance characteristic that is not included in the manufacturer's specifications information, although it could be measured scientifically. Distance sensitivity is one that must be learned from experience and must be anticipated for the individual environment and recording conditions.

Many variables must be considered during the process of selecting and placing microphones. The primary variables for microphone selection were presented above. The variables of *microphone placement* will directly influence the selection of a microphone, even after an initial selection has been made. Placing the microphone in relation to the sound source and the performance environment will greatly influence the sound quality of the recording. At times, this influence may be as great as the microphone itself.

The recordist must consider the following when deciding on the location of the microphone, in relation to the sound source:

- Distance relationships between the microphone, the sound source, and the reflective surfaces of the environment;
- Distance of the microphone from the sound source to be recorded, and other sound sources in the environment;
- Height and lateral position of the microphone in relation to the sound source;
- Angle of the microphone's axis to the sound source; and
- Performance characteristics of the microphone selected, as altered by the previous four considerations.

Microphone placement interacts with microphone performance characteristics, to create the recorded sound quality of the source. The distance of the microphone will be largely determined by the distance sensitivity of the microphone. The angle of the microphone will be largely determined by the frequency response of the microphone in relation to its polar pattern, the characteristics of how the sound source projects its sound, and the characteristics of the environment. The height of the microphone is also a result of the characteristics of the sound source, the environment, and the microphone's frequency response and distance sensitivity.

The sound characteristics of the environment in which the sound source is performing may or may not be captured in recording the sound source. The recordist may or may not wish to include them in the sound quality of the sound source. In either instance, the recordist must be in control of the indirect sound of the environment arriving at the microphone from reflective surfaces.

The recordist may seek to control the balance of direct and indirect sound. Through selecting and placing a microphone with a suitable polar pattern and distance sensitivity, the desired characteristics of the environment may be captured, in the desired amount, along with the sound of the sound source. This will allow the recordist to record the sound of the sound source within its performance environment. The distance cues of the initial reflections in the early time field and the amount of timbre detail are evaluated by the recordist, when deciding on the ratio of direct to indirect sound, and distance placement is adjusted accordingly.

The recordist's objective may be to capture the sound source without the cues of the environment. The sound source may be physically isolated (with gobos or in isolation booths) from other, unwanted sounds from the environment and other sources, or the leakage of unwanted sounds to the recording microphone might be minimized with microphone pattern selection and microphone placement. This will allow the recordist the flexibility of being in complete control of the sound source. Environmental characteristics are later applied to the sound through signal processing. The sound of the environment will be controlled by the relationships of the microphone to the sound source, and any other sound sources that may be occurring simultaneously, including the sound of the environment itself.

The *reflective surfaces* of the environment (or of any object in the environment) can cause the sound at the microphone to be unusable. Interference problems may be created when the sound from a reflective surface and the direct sound reach the microphone at comparable amplitudes. The slight time delay between the two sounds will cause certain frequencies to be out-of-phase (with cancellation of those frequencies) and certain frequencies to be in-phase (with reinforcement of those frequencies).

The frequencies that will be accentuated and attenuated can be determined by the difference of the distance between the reflective surface and the microphone (D1) and the distance between the sound source and the microphone (D2), in relationship to air velocity. The amount of reinforcing and cancelling of certain frequencies that will be perceived when the two signals are combined will be determined by their particular amplitudes. Constructive and destructive interference are most pronounced when the two signals are of equal amplitudes. As the difference in amplitude values between the two signals becomes larger, the effect becomes less noticeable.

The constructive and destructive interference of the combined signals result in a frequency response, with emphasized and attenuated frequencies (peaks and dips). These peaks and dips of the frequency response curve of the sound's spectrum have been compared, in analogy, to the tines of a comb. Thus, the term *comb filter* has been applied when a signal is combined with itself, with a slight time delay between the two signals.

The distance of the microphone to the sound source and to other sound sources in the environment, along with the height, lateral position, and angle of the microphone, alter the sound quality of the sound source. As such, they are variables that may be used as creative elements in shaping sound quality (and the musical material), in the same way as microphone selection, discussed earlier. These areas will most significantly influence the following aspects of sound:

- Environmental characteristics (secondary distance) cue: ratio of direct to reflected sound;
- Environmental characteristics cue: early time field information created by initial reflections;

- Distance cue: definition or amount of timbre detail; and
- Capturing of the blend of the source's timbre.

The distance from the microphone to the sound source is the primary factor that controls the ratio of direct to reflected (indirect) sound. As the microphone is placed closer to the sound source, the proportion of direct sound increases in relation to reflected sound.

Reflective surfaces that are located near the sound source will cause inconsistencies with this general rule. If at all possible, sound sources should be placed at least three times the distance from any reflective surface, as the distance from the microphone to the sound source. The height and angle of the microphone in relation to the sound source, coupled with polar patterns, may allow the microphone to keep from picking up the reflected sound off surfaces close to the sound source (such as the floor or a gobo).

Distance cues are established by microphone placement when an audible amount of sound from the recording environment is present in the source's sound quality. While this is not usually the case with close miked sound sources, many sound sources in multitrack projects are recorded from a distance that will capture certain information from the recording environment. Of course, all distant micing techniques will capture a significant amount of the sound of the environment.

The reflection information in the early time field is especially prominent in recordings that have utilized a moderate distance between the microphone and the sound source. These reflections will often be caused by the floor or other objects immediately around the sound source. While only a few reflections may be present in the final sound, and the reflections may be of significantly lower amplitude than the direct sound, they will impart important environmental characteristics information. These reflections often provide environmental cues that conflict with the perceived stage-to-listener position distance cues.

Microphone placement location and performance characteristics will play vital roles in establishing distance cues, through defining the amount of *timbre detail* present in the recorded sound source.

Generally, the closer the microphone to the sound source and/or the more sensitive the microphone in terms of distance sensitivity and transient response, the more timbre detail that is captured when recording the sound source. This provides a distance cue that may place the listener of the recorded sound near the reproduced image of the recording. This may be accomplished in such a way that the listener may imagine himself or herself performing the instrument, located beside the performer of the instrument, or even located adjacent to the location where the instrument is sounding (the location of the recording microphone).

Distance cues are often contradictory when a high degree of timbre detail and a large amount of reverberant energy in relation to direct sound are present simultaneously. This unrealistic sound may be desired for the artistic process, but the recordist must be aware of this dichotomy in order to be in control of the medium.

The sound quality of the sound source may potentially be altered by close microphone placements. As stated earlier, these alterations can be used to creative advantage, or they can distort the desired sound quality. The sound quality may exhibit the following alterations, when the source is close miked:

- An increase of definition over known, naturally occurring degrees of timbre detail;
- Changed spectral content of the sound source; and
- An unnatural blend of the source's timbre, caused when the source has not had enough physical space to develop into its characteristic sound quality.

A microphone placed within two feet of a sound source may alter the source's spectral content. The frequency response of certain microphones are altered by the *proximity effect*. Response in the low frequency range rises relative to response in higher frequencies, for cardioid and bidirectional microphones.

The sound source itself and the way it radiates sound, along with the producer's idea of what it should sound like, must also be considered in the context of the sound's environment.

Sound sources require physical space for their sound quality to develop or coalesce. The sound quality of instruments and voices is a combination of all the sounds the instrument produces. When a microphone is placed in a physical location near the sound source, it will often be within this critical distance necessary for the sound to develop into a single sound wave. The sound source will not have had the opportunity to *blend* into its unique, overall sound quality. Only a portion of the sound will be captured by the microphone. This sound quality will be very different from people's experiences of how the sound source appears in acoustic environments.

It is not unusual for recordists to capture a sound before its characteristic acoustic sound has fully developed. This can be accomplished in such a way as to contribute to the music, or it may work against the project. It is important that the recordist be aware of this space and be in control of recording the desired blend of the source's sound.

Many instruments and voices are commonly recorded with close microphone techniques, capturing only a portion of the source's characteristic timbre. Care must be taken to obtain the desired sound quality; many undesirable noises are created by performers and their instruments that may cause difficulties in obtaining a suitable sound.

In close micing sound sources, all or most natural environmental cues are absent from the sound. Environmental characteristics will be added to these sources through signal processing. Electronic or amplified instruments that are plugged directly into the mixing console (through a direct box) are similar in their lack of environmental characteristics. If the listener is to be provided with a realistic listening experience (which may or may not be desirable), environmental characteristics need to be added to these sounds.

Recordists often capture the sound of a sound source from a number of close microphone locations, to blend the sound themselves. This can be performed to great creative advantage; the recordist may control the blend of the various portions of the source's sound, instead of relying on the sound source's interaction with the performance environment to blend the sound as desired. Pianos are commonly recorded in this way, as are acoustic guitars. This technique can be applied to a wide variety of sound sources.

Sound sources (individual instruments/voices or ensembles) may go through a conceptual *preprocessing*, with the application of microphone techniques. Preprocessing is an alteration to the components of the sound source before it has become available in the routing and mixing stages of the recording chain.

A stereo microphone array uses two or more microphones, in a systematic arrangement. It is designed to capture the sound in such a way that upon playback of the recording, through two loudspeakers, an accurate sense of the spatial relationships of the sound sources present during the recorded performance is reproduced. The techniques attempt to capture the sound qualities of the live performance and of the hall, the performed balance of the instruments of the ensemble, and the spatial relationships of the ensemble (stereo location and distance cues), with minimal distortion.

Stereo microphone techniques are often used in recording large ensembles, in large acoustic spaces, from a rather distant placement. The techniques are very powerful in their accuracy and flexibility and may be applied to either a single sound source or to any sized ensemble. They may be used from a rather distant placement (perhaps 15 or more meters, depending on the pertinent variables), to within about 1 meter of the sound source (stereo arrays are commonly applied to drum sets–sometimes supplemented with accent microphones, sometimes not).

The stereo microphone array can significantly alter the sound source in a number of ways. All stereo microphone techniques have their own unique characteristics and, as such, their own inherent strengths and weaknesses. The inherent sound qualities of the stereo microphone arrays can be used to great advantage if the recordist understands and is in control of their sound qualities. The recordist will need to become aware of the sound qualities of each of the various arrays, evaluating their abilities to function according to the following list of variables, in the context of the above mentioned microphone placement variables:

- Perceived listener to sound stage distance;
- Amount and sound quality of the environmental characteristics of the performance space;
- Perceived depth of the ensemble, sound source, or sound stage;
- Perceived width of the sound source or the sound stage;
- Definition and stability of the lateral (stereo) imaging;
- Musical balance of the sound sources in the ensemble; and
- Sound qualities of the entire ensemble or of specific sources within the ensemble.

The recordist will often use the concept of the ideal seat when determining the placement of a stereo microphone array. The placement of the array will theoretically provide sound qualities that are the most desirable for the particular ensemble performing a specific piece of music, in the most appropriate performance space. The ideal seat is related to the concept of creating a perfect performance; the recordist will seek to balance the hall sound with that of the ensemble, capture an appropriate amount of timbre definition from the ensemble, retain all performed dynamic relationships, and establish desirable and stable spatial relationships in the sound stage.

Many stereo microphone techniques have been developed. Among the most commonly used are:

- X-Y coincident techniques;
- Middle-Side technique (M-S);
- Blumlein Pair (crossed figure-eights);
- Near coincident techniques (NOS and ORTF);
- Spaced omnidirectional microphones;
- Spaced bidirectional microphones;
- Binaural system (artificial head); and
- Sound field microphone system.

Accent microphones are often used to supplement the stereo microphone techniques discussed earlier (with the exception of the specialized binaural and sound field recording systems). These are microphones that are dedicated to capturing a single sound source, or a small group of sound sources, within the total ensemble being recorded by the array. The accent microphones are placed much closer to the sound sources than the array, and may cause the recordist to consider some of the close micing variables discussed above.

At times, secondary stereo arrays can suitably function as a set of accent microphones. This is especially usable with large ensembles (such as an orchestra plus chorus). The recordist must be aware of delay times between the arrivals of the accent array(s) and the overall array signals. These time differences may be minimized by applying delay units to the secondary array(s), thus avoiding comb filtering distortions.

When using accent microphones in conjunction with stereo microphone arrays, the accent microphones are most often used to complement the array. They assist the overall array by bringing more dynamic presence and timbre definition to certain sound sources in the ensemble, and they allow the recordist some control over the musical balance of the ensemble. Accent microphones also create noticeable time differences between the arrival of the sound source(s) at the stereo array and the arrival of the sound source(s) at the accent microphone.

Adding accent microphones will diminish the realistic nature of the stereo microphone array, in accurately capturing the performance (while attempting to improve the sound present in the hall). Accent microphones will alter time cues and dynamic relationships noticeably, making the sound source(s) captured by the accent microphone significantly more prominent in the musical texture.

EQUIPMENT SELECTION: APPLICATION OF INHERENT SOUND QUALITY

With a clear understanding of the desired sound qualities of the individual sound sources and the entire recording project, the recordist must determine which pieces of the equipment will be utilized to achieve the desired result. The recordist will approach this problem by evaluating the inherent sound characteristics of the available individual devices and the inherent sound characteristics of the technologies of those devices against the unique needs of the individual project.

Digital recording, processing, and editing equipment is not necessarily "better" than analog equipment, in relation to its potentials for artistic expression, nor is the opposite true. No technology is inherently better suited than another for generating, capturing, shaping, mixing, processing, combining, or recording sound.

Analog technology has certain inherent sound characteristics. Digital technology has certain inherent sound characteristics. The characteristics of one technology may or may not be appropriate for the particular project. Inherent sound qualities are inherent sound deficiencies if they work against the sound quality that the recordist is trying to obtain. Inherent sound qualities are desirable if they produce the desired sound quality for the recording.

It is difficult to make generalizations as to the characteristics of analog versus digital technology. The sound qualities of both technologies vary widely, depending on the particular unit and the integrity of the audio signal within the particular devices. An 8-bit digital system is significantly less accurate and flexible than a 32-bit system. A consumer-model analog system is significantly more noisy and less accurate than a professional unit.

Differences often exist between the two technologies in:

1. Their abilities to accurately track the shape of the waveform (especially the initial transients of the sound wave, in both technologies);
2. Their abilities to process all frequencies equally well (especially frequency response problems of analog, or quantization errors in digital);
3. Their ability to store the waveform, without distortion from the medium (especially tape hiss or A/D and D/A conversion errors);
4. Their ability to alter the waveform in precise increments and to precisely repeat these functions (a measure of signal processors); and
5. Their ability to perform repeated playings, successive generations of copying, and long-term storage, with minimal signal degradation (a measure of recording formats).

Many other, more subtle, differences exist, especially between specific devices of each technology.

Recordists will develop sound quality preferences and working preferences for particular pieces of equipment and/or technologies. Developing such preferences may or may not be artistically healthy. A recordist may become inclined to consider a certain technology to be "better" than another because it is the one he or she is most familiar with, not because it is the one that is most appropriate for the project. Often, personal preferences (or personal experiences) are confused with the actual quality or usefulness of a device or a technology.

Personal preferences are not a measure of quality, in and of themselves. Audio devices have inherent sound qualities that are largely determined by technology. The recordist will be using these devices to shape the sound of the music recording, and will do well to learn the inherent sound qualities of as many devices as possible. These are the tools (instruments) that will be used in *The Art of Recording*.

Similarly, no recording device is inherently better than another similar device. While certain devices are more flexible than others, and certain devices are of higher technical quality than others, the true measure of a piece of equipment is its suitability to the particular needs of the project, at a given point in time.

The advantages of any particular device in one application may be a disadvantage in another. The sound quality of one device may be appropriate to one musical context and not to another. The measure of the device will be in how its inherent sound qualities compare to the desired sound qualities of the recording. Pieces of recording equipment will be evaluated by the recordist, for their sound qualities and their potential usefulness in, communicating the artistic message in the piece of music. This evaluation is performed through a critical listening process similar to that used to evaluate microphone performance characteristics.

Pieces of recording equipment are tools. The tools may be applied to any task, with consistent results. The recordist needs to decide if a particular tool (piece of equipment) is appropriate for the sound quality that is required of the particular project.

Musicians often carry with them a number of musical instruments. They will use a different model of the same instrument (perhaps made by a different manufacturer) to obtain a different sound quality of their performance, depending on what is required by the musical material. The recordist should recognize that this is similar to their own circumstances.

In selecting recording equipment, the recordist is, in essence, selecting a musical instrument. The sound quality of sound sources or of the entire recording may be markedly interpreted by the piece of equipment, while the sound is under the control of the recordist. This is the way a traditional musical instrument is applied by a traditional performer.

Among the most commonly applied audio devices are signal processors. They comprise a large portion of the recordist's instruments, or tools and may play a significant role in shaping the individual project. Specific devices are chosen because their individual, inherent sound qualities lend themselves to the particular project.

The sound quality of many processing devices may be altered to sound similar to very different devices, merely by how they are applied to the sound source and how they are used in conjunction with other processors (stringing processors together on a single sound source). If the recordist is knowledgeable of how the devices operate on the signal and the physical components of the sound source, he or she is able to achieve a remarkable number of different sound qualities out of a small number of processors.

The primary differences between models of equipment are the flexibility with which they can modify the signal and the integrity or technical quality of the signal after the processor has been applied. Less expensive processors often distort or add noise to the signal; these devices may not be suitable for many professional applications. The device should allow the recordist very precise control in modifying the physical components of the signal. Devices that store settings that the recordist is tuning for applications in a project are very helpful, although these are found almost exclusively on digital devices.

The feature offered by a specific device, or a certain technology, may be a factor when deciding on one piece of equipment over another. A piece of equipment that is capable of recalling settings that the recordist has established many weeks before, perhaps at a different studio, from a floppy disk or from a data cartridge may make one device more desirable than a similar, analog device. Conversely, the recordist's detailed documentation of processor settings and applications may be used as a very accurate and reliable substitute for data storage and recall. The selection of one device over another is not always a clear, sequential process.

MONITORING: EXTRANEOUS SOUND QUALITIES AND THE LISTENING EXPERIENCE

During the preliminary stages, the recordist must decide on how to approach the perception (monitoring) of the recording process. Monitoring the recording process in the recording control room (whether it be a commercial facility or a closet used for a remote recording) will be the means through which the recording itself will be evaluated. All the sounds and relationships of the recording project are presented to the listener through the monitoring system.

The monitor system is more than a pair of loudspeakers. In terms of hardware, the monitor system also includes power amplifiers, crossover networks, and (perhaps) the monitor mixer of the console. All hardware must function efficiently and effectively in relation to the other components, in a complementary manner in terms of sound qualities, to reproduce a waveform that has not been distorted. The monitor system also encompasses the listening room itself, the placement of the loudspeakers in the room, and the interactions of the room and the loudspeakers (especially reverberant energy and strong reflections).

The monitor system has the potential to distort all sound qualities and relationships in the recording. This must be avoided as much as possible, or at the very least, the recordist must know how the sound is being distorted. The recordist must evaluate the project from many different perspectives, focusing on any artistic element. Each of these perspectives will require the monitoring system to accurately reproduce sound quality on all hierarchical levels, the spatial relationships of sound sources, and the frequency, amplitude, and time information of the recording.

These evaluations must be of the recording itself, not of the recording through the monitor system, if the recordist is to be in control of the recording process and if the recording is to have the same sound relationships in another (neutral) listening environment.

The most desirable monitoring system is transparent; the loudspeakers reproduce the sound in the control room, without distorting the sound quality. This seems to be almost an impossibility. Many complexities are present in the selection of an approach to monitoring. The immediate considerations for the monitoring process itself are:

- Loudspeaker and control room interaction;
- Effective listening zone;
- Sound field: near field versus room monitoring;
- Monitoring levels; and
- Stereo/mono compatibility.

The *control room* itself will distort the sound emanating from the loudspeakers. The acoustics of the control room can cause radical changes in the frequency response and time information of the sound. Ideally, any listening room would have a constant acoustical absorption over the operating range of the loudspeakers and would appropriately diffuse the sound from the loudspeakers, to create a desirable blend of direct and reflected sound at the mixing/listening position.

The influence of the control room on the sound coming from the loudspeakers will be minimized by:

- Nonparallel walls;
- Nonparallel floor and ceiling;
- Acoustical treatments to absorb, reflect, and diffuse sound where needed;
- Carefully selecting, placing, and installing loudspeakers; and
- Using a room with sufficient volume (dimensions of the room).

The room should absorb and reflect all frequencies equally well and should produce very short decay times that are at substantially lower levels than the direct sound from the loudspeakers. The room should not produce resonance frequencies and should not produce reflections that arrive at the listening position at a similar amplitude as, or within a small time window (2 to 5 ms) of, the direct sound.

As needed, rooms may be tuned for uniform amplitudes of all frequencies by room equalizers and tuned for controlling reflections (time) with diffusers, traps, and sound absorption materials. The ideal control room has been designed to include very specific acoustical treatments, in such a way that room equalization is not needed.

Designers of recording studios have widely divergent opinions on the most desirable acoustical properties of control rooms. Conflicting information and opinions are common. The objective of all designers is very similar, however: produce a listening environment that is most suitable for the listening of reproduced sound.

Loudspeaker placement is part of the design of control rooms. Ideally, the loudspeaker should be mounted within the wall so that the front of the loudspeaker is flush with the face of the wall. The speakers become part of the wall, negating the usual boost of low frequencies that results when a loudspeaker is placed near walls, ceiling, and floors (and especially in corners). Placement away from side walls by four feet and from the front wall by about three feet will minimize the boost of low frequencies that occurs when the omnidirectional low frequencies reflect off the wall surfaces and combine with the direct sound from the speaker.

The loudspeakers should be aligned on the same vertical plane, usually at ear height, for a seated person, or slightly higher. It is important that the meter bridge of the console is not in the path between the loudspeakers and the mix position, and that strong reflections off the console are minimized. The loudspeakers should be aligned symmetrically with the side walls.

The *effective listening zone* is that area where the sound is accurately perceived, as it is reproduced through the two loudspeakers interacting with the control room. In a control room, the mix position and/or the producer's seat are located in the effective listening zone. The size of the effective listening zone is usually quite small in most control rooms; it is an area that is equidistant from the two loudspeakers and roughly the same distance from each speaker as the speakers are from each other. Few strong reflections arrive at this area. Angling the loudspeakers (from 60 degrees with respect to the listener, to 90 degrees with respect to each other) towards the listening zone increases the size of the effective listening zone, given complementary room acoustics.

For accurate spatial perception, it is vital that the effective listening zone be carefully evaluated. The listener must be located almost exactly between the two loudspeakers (depending on the acoustics of the control room), and the volume level must be the same at each speaker when identical signals are applied to each channel. The control room should be virtually transparent in the listening zone (Fig. 5-2).

The effective listening zone of the control room is in relation to the *studio monitor speakers*. This is the monitoring system that provides the most accurate information of the qualities of the recording being made. The studio-quality loudspeakers are of high quality and are carefully engineered. They are designed to provide non-distorted, detailed sound, while working within the acoustic characteristics of the control room.

Most people who purchase the recording will have playback (monitoring)

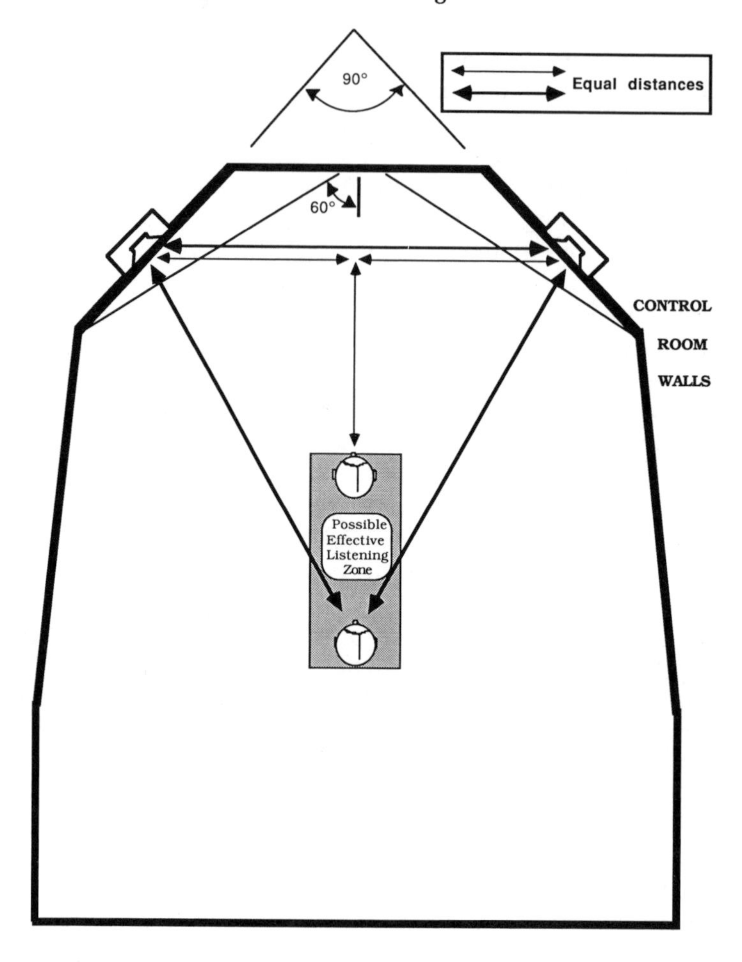

FIGURE 5-2. Loudspeakers as Part of the Control Room, and the Effective Listening Zone.

systems of somewhat (to significantly) less quality than the recording studio. The sound of the recording in the audience's listening environment (the individual's home, automobile, etc.) will be different from that of the recording control room. It is common for recordists to use several sets of monitor speakers to obtain an idea of how their project will sound over home-quality playback systems.

Bookshelf-type speakers are often used in the studio to provide a reference to this type of playback environment. They represent the typical, moderately priced speaker systems of most home listeners' rooms. They tend to have a narrow frequency response, de-emphasizing high and low frequency areas. These speakers are placed on the meter bridge of the console, about three to five feet apart. This is called *near field* monitoring. The sound is heard near the speakers, before the acoustics of the control room can act upon the sound.

Since the control room has very little influence on the sound quality of near field monitoring, this approach is often the exclusive monitoring system of control rooms that have poor room acoustics. Near field monitoring can be more accurate than many home systems. Certain studio monitor speakers have been specifically designed for near field use. These speakers are preferred over the large studio monitors by some recordists, especially during long sessions when the more intense energy of some large speakers may fatigue the hearing.

The size of the effective listening zone for near field monitoring is very small (Fig. 5-3). The listener must be precisely centered between the two loudspeakers and cannot be further from the two loudspeakers than they are from one another, or the effects of the room will come into play.

It is not unusual for recordists to periodically switch back and forth between room and near field monitors. This allows the recordist to evaluate the sound qualities and relationships of the recording from different listening locations and with loudspeakers that have different sound qualities. The recordist will be attempting to create a consistency between the

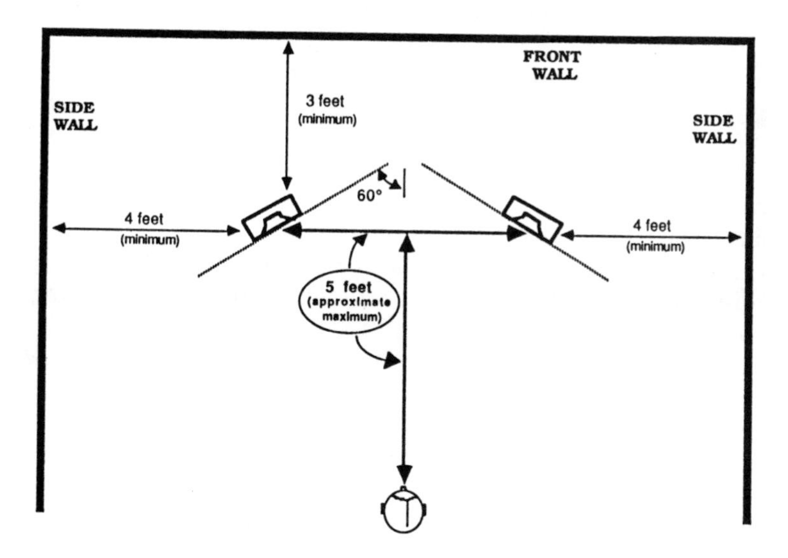

FIGURE 5-3. Loudspeaker Relationships to the Listening Environment and the Listener in Near Field Monitoring.

sounds of the two monitor systems, as much as possible. The speakers should be switchable at the mixing position, with each pair having a dedicated amplifier(s), set to match monitoring level while switching speakers. Control room monitors tend to provide much more detail of sound quality than near field monitors. This added detail allows for greater accuracy in critical listening applications.

The sound pressure level (SPL) at which the recordist listens will influence the sound of the project. Humans do not hear frequency equally well at all amplitudes. A recording created at a *monitor level* of 100 dB (SPL) will sound considerably different when it is played back at the common, home listening, monitoring level of 75 dB; the bass line that was present during the mixdown session will not be as prominent during the listening at the lower level. Similarly, the tracks that were recorded at 105 dB during a tracking session (because someone wanted to "feel" the sound) will have quite different spectral content when they are heard the next day at 85 dB, during the mixdown session. The recordist will need to develop consistent listening levels. Ideally, monitoring levels should be reasonably consistent throughout a project.

The most desirable range for monitoring is 85 to 90 dB. Recordings made while being monitored in this range will exhibit minimal changes in frequency responses (≤5 dB at the extremes of the hearing range) when played back as low as 60 dB SPL. Monitoring at this level will also do much to minimize listening fatigue during prolonged listening periods (recording and mixing sessions).

The possibility of hearing damage is a very real job hazard for recordists. If the recordist primarily listens in the 85 to 90 dB range, he or she will have accurate hearing much longer than if monitoring is done at a level 10 to 15 dB higher.

The recordist may need to be sensitive to how the recording being produced will transfer to other environments and playback formats. Aside from the varying quality of home playback systems described above, the recordist may need to be concerned about the playback of a recording over radio and in a mono format.

The targeted audience for some recordings may necessitate that the recordist be concerned with the *stereo/mono compatibility* of the recording. The recordist will monitor for this format, using a single-driver mini-monitor speaker, in the near field. This type of speaker is intended to simulate the listening conditions of a radio or an automobile playback system. The recordist may determine the mono compatibility of the stereo program (being mixed on the studio monitors) by routing the monitor signal to the single loudspeaker or by summing the two channels through the console's "mono" switch and using both of the studio's usual monitor speakers.

When the two channels are summed, conflicting phase information causes comb filtering and other signal-integrity problems. In addition, the balance of the sound sources can be radically altered; center-image sound sources may be raised as much as 3 dB, when the signals of the two channels are combined. For these reasons, it is not uncommon for separate mixes of the same work to be made for different releases. Mixes made for radio broadcast will often have altered frequency information to compensate somewhat for the inadequacies of the single, small speaker, and will seek to maintain many of the dynamic level relationships achieved in the stereo mix.

Headphones are required for creating and listening to binaural recordings, and may be the only feasible monitoring option for a remote recording. Otherwise, monitoring is most accurately accomplished over loudspeakers, except in the rarest of circumstances.

Headphones will distort spatial information, especially stereo imaging. The listener will perceive sound within their head, instead of as occurring in front of their location, which distorts depth of imaging. The interaural information of the recording is not accurately perceived during headphone monitoring, causing potentially pronounced image size and stereo location distortions.

Further, the frequency response of headphones will not match that of loudspeakers, and it is inconsistent. Frequency response of headphones will vary with the pressure of the headphones against the head, and the resulting nearness of the transducer to the ear.

Sound evaluations can only be accurately performed over quality loudspeakers, in a transparent (or complementary) listening environment.

6

Capturing, Shaping, and Creating the Performance

The recordist will capture the performance through tracking the musical parts and instrumentation, and will shape sound qualities and relationships of the performance through signal processing. The mixing process will continue to shape the performance and will result in the creation of the performance (that is, the recording).

With tracking, the actual act of recording sound begins. The preliminary stages that were discussed in the previous chapter define the dimensions of the creative project and of the final recording. In addition to these, certain other preparations must take place before a successful tracking or recording session can begin.

Whenever possible, the musicians should be prepared for their performance at the tracking session. Unfortunately, it is the norm for musicians to get their music when they arrive for a recording session. This limits their potential creative contribution to the project, and may limit the quality of their performance.

Ideally, the music should be given to them well in advance of the session, they should have the opportunity to learn their parts before they join the other musicians, and all groups of musicians that will be performing simultaneously should be well rehearsed together (as an ensemble). As much as possible, musicians should be given the opportunity to arrive at the session knowing what will be expected of them, and ready to perform. In the reality of music recording, this ideal rarely occurs.

Some final instrumentation decisions and many of the supportive aspects of the music are, however, often decided upon during the preliminary stages of the recording process (or during the production process itself). The recordist should anticipate that many musical decisions will be made during tracking; he or she should be aware that the dimensions of creative projects tend to shift (sometimes markedly) as works evolve. The recordist must keep as many options open as possible; this keeps the creative artists from being limited in exploring their musical ideas. The recordist must seek to always be in the position of being able to execute any musical/production idea requested of them, without hours of undoing or redoing tracks (which may often be caused by a production decision made and executed by the recordist a few minutes before the musical/production idea was conceived). Allowing for flexibility may be as simple as leaving open the option of adding more instruments or musical ideas to the piece, by leaving open tracks on the multitrack tape, in console layout, cue mix changes, signal routing, or in mixdown planning.

It has often been said that the recordist should first be a psychologist. While this statement may be somewhat extreme, the recordist must be aware of interpersonal relations. The ways that people normally treat one another in everyday living, and especially in standard business environments, are often nonproductive (at best) in the recording studio. Recordists should consider how they speak with the artists about the project and how they interact with the artists socially and during the creative processes. The type of image the recordist presents to the artists will influence their ability to function well professionally and will keep the artist relaxed about their environment and focused on the project.

Recordists are trying to get creative people to do their best work, while attempting to perform their own tasks at the highest standard. The process of creating art (a music recording) is an emotional roller coaster of ecstasy of what has just been discovered, and anguish over not having an equally brilliant answer to "what comes next?" The creative process is further stressed by the artists' time and financial constraints in the studio.

Musicians and creative people are at their most exposed and vulnerable state during the recording process. The recordist must be certain to do nothing (or to allow anything to happen within the session environment, or by anyone else) to make the artists feel unprotected or, worse, threatened. The musicians must feel that they have the freedom to be creative around the recording studio, without being constantly judged or evaluated; at times, they will need to feel they are alone. Evaluations have their place in the recording process; critical judgements can rarely be used constructively by the recordist.

The recordist must know when to make evaluations and what to evaluate, *if* it is within their function for the particular project. The expressive nature of performing music will often involve taking chances, stretching performance abilities to their limits, and making mistakes. These necessary activities can potentially embarrass confident, let alone less-than-confident, performers if they are critically judged at this vulnerable time. Performers need to be confident to perform well, and the recordist is attempting to get the performers to play at the height of their ability. Nothing should be allowed to happen that would diminish the confidence level of the performers, or take away from the trust that the performers must have in the recordist.

The recordist must not interfere with the creative process, yet he or she is often responsible for keeping it moving effectively, efficiently, and invisibly. The artist must be free to be creative and completely involved in the project; all things should make the artist feel secure, relaxed, and focused, and to make the session productive. What the recordist talks about, the tone of voice, and the language selected all influence how the artist feels. The assertiveness of the recordist's personality should be adjusted appropriately for the situation of the particular artist and the particular project.

All distractions from the recording process are kept to a minimum. Talk is to be centered around the project, with rare exceptions. Breaks and social conversation are most effective out of the space where work occurs.

The recordist must know his or her function. If unclear about their role, he or she should have it clarified quickly by the person in charge. By being sensitive to the needs and preferences of the artists/producer, and through anticipating them without being noticeable, the recordist will greatly assist the progress of the session/project. The recordist should attempt to be transparent and unnoticed, yet be ready to be as helpful as required. An atmosphere conducive to relaxed, creative, productive work must be presented. This will make the artist forget about the time element, yet keep the session moving in a clear and organized manner.

The recordist should not offer an evaluation unless it is asked for by someone who really does want an honest answer, or unless they are contributors to the creative process. The recordist's evaluation must be objective to be of use. The personal taste, life experiences, and feelings of the recordist are not what has been requested. Provide an honest (and sensitive) evaluation of the material in relation to the context of the session and the function of the project. The recordist must not be assertive when another person is responsible for the context and quality of the project, except under circumstances where intervention will avoid a catastrophe.

The recordist's own creative ideas come into play when he or she is directly involved with the creative decisions of the project. At that point, the recordist is a creative artist. The interpersonal dynamics of the group will shift from the above; the recordist is an equal member of the group and often leads the group. The recordist's position in the creative process can be a very powerful one in capturing, shaping, and creating the music recording. He or she will be a performer or the creator of the music. This is the recordist as an artist.

The recording studio is the musical instrument of the recordist; the recording process is the musical performance of the recordist. In order to use recording for artistic expression, the recordist must be in complete control of the devices in the studio, and must understand their potentials in capturing and altering the artistic elements of sound.

TRACKING: SHIFTING FOCUS AND PERSPECTIVES

Tracking is the recording of the individual instruments or voices (sound sources), or groups of instruments or voices, onto a multitrack tape. This is done in such a way that the sounds can be mixed, processed, edited, or otherwise altered at some future time, without altering the other sound sources on the tape. It is imperative that the sound sources are recorded with minimal information from other sound sources and at the highest safe loudness level that the system will allow, if this isolated control of the sound source is to be possible.

The tracking process must be well planned. The number of available tracks for the recording project is the major limiting factor. Within this limitation, the recordist must plan track assignments, (the sounds to be placed on each track of the tape). These include:

- All anticipated instruments will have at least one track assigned to them, with several tracks left unassigned, if possible.
- Specific track assignments may be required for specific instruments.
- Certain sources may need two tracks for spatial characteristics.
- Tracks that should remain unused to assist in bouncing materials and in compiling submixes (as necessary).
- Tracks that may be dedicated to more than one sound source (when certain instruments are only present during specific portions of the piece).
- Some tracks may need to be dedicated to sync tones or automation data, and will require special handling.
- Suitable location of reference tracks (unpolished performances that are used as a guide for the other musicians, especially during the early stages of tracking), where they can be erased and of use at later production stages.

The order of the performances of the tracking session should be well planned, to coincide with track utilization and (most importantly) effective use of the musician's time and concentrated, creative energies.

Tracking requires the recordist to continually shift focus and perspective, and to move between analytical and critical listening processes. The musical qualities of the performance must be evaluated at the same time as the perceived qualities of the captured sound. The recordist is responsible for making certain that both aspects are of the highest quality and exist in a suitable state on tape.

Focus will have the listener's attention moving freely between all the artistic elements of sound and all the perceived parameters of sound. The recordist will often be required to shift focus between artistic elements (shifting their attention between dynamic levels, pitch information, or spatial cues) and then immediately to a focus that alternates between the perceived parameters (such as timbre definition). The recordist is required to continually scan the sound materials, to determine the appropriateness of the sound (and its artistic elements) to the creative objectives of the project, and to determine the technical quality of the perceived characteristics of the sound.

The ability to shift focus must be developed in order to allow the recordist to function. He or she must be aware of the musical and technical qualities of the sound being produced, and be in complete control of the recording process, so as to shape the sound as desired. Tracking most directly utilizes the recordist's ability in shifting focus and perspective. Although these skills are used in all other aspects of recording, successful tracking cannot be effectively performed unless the recordist is able to quickly and effectively recognize and address all the states of sound, as they function simultaneously.

Perspective will be shifted often during the tracking process. In critical listening processes, the recordist will shift perspectives of detail, to seek information on the integrity of the sound being recorded. These levels may include listening from the perspective of the activities of the most minute aspects of a sound source's timbre as it is captured by a microphone, or listening to the timbre of the monitor mix as it is influenced by the playback system, to give two illustrations.

Perspective will be used in the analytical listening processes to evaluate the artistic elements and how they are being utilized in the recording process. The recordist will move between levels of detail that are related to the creative aspects of the project. The process involves scanning the sound information, at various levels of detail, to determine that the sound materials are in their intended states and that they are progressing as desired.

Within the tracking process, the recordist is concerned about the quality of the performance of what is being placed on tape, and about the technical quality of what is being placed on tape. Depending on the particular project, he or she may not be in the position to make decisions related to *performance quality*, but should nonetheless be aware of the performance and be ready to provide evaluations when asked. It is imperative that the recordist always be aware of the *technical quality* of the recording; sound must be recorded and reproduced, without being distorted.

There are many performance quality aspects of music. Most of these aspects are related to intonation, control of dynamics, accuracy of rhythm, and performance technique. The performance technique of performers is critical in recording, as their sound cannot be covered up by the other players or assisted by the acoustics of a performance environment. The ways that performers produce sound on their instruments can create the desired musical impact, or it may not. The person responsible for the project will need to make evaluations and offer necessary alternatives.

The performance technique problems and the natural acoustic sound properties of all of the instruments (well outside the scope of this writing) should be known to the recordist. Just as the deductive selection of a microphone to accurately capture the sound of the instrument relies on a knowledge of the sound source, negating performance technique difficulties (when necessary) relies on a detailed knowledge of the sound source, along with a knowledge of the capabilities of the various components of the recording chain.

Performance intensity information and all other expressive qualities are vital to the music. Performance intensity becomes an integral part of sound quality in recordings. The dichotomy between perceived dynamic levels and perceived performance intensity is often utilized to creative advantage in music recordings. The producer (recordist) should give particular attention to performance intensity and expression, as sound quality resources, during tracking.

Final decisions on the sound quality of the sound sources are often delayed to the mixdown stage. Some shaping of sound quality does take place during the tracking stages, as discussed below.

The technical quality of the recording will be reflected in the integrity of the recording's sound qualities. The recording process must not be allowed to alter the perceived parameters of sound, unless there is a particular artistic purpose or technical function for the alterations. The perceived pitch, amplitude, time elements, timbre, and spatial qualities of the recorded sound sources should be accurately captured in the recording and should be the only sounds present in the reproduced recording. Any extra sounds are noise; any unwanted alterations to the sound are distortions. The audio signal must be accurately recorded and reproduced, for artistic expression to be accomplished in a controlled manner. Changes of focus and perspective during the critical listening process will make this possible.

Other critical listening applications during tracking are centered around performing evaluations of the sound quality of sound source, as they have been altered by the creative applications of the recording process. Among the most common concerns are:

- Sound sources isolated during tracking;
- Undesirable sound quality of the sound source (often caused by microphone selection or placement); and

• Unwanted sounds created by performers or instruments.

Sounds will have a particular amount of isolation from one another. If the sounds are to be altered individually in the mixdown stage, they must be isolated. When a group of sounds function as a single unit, it may not be necessary (or appropriate) to isolate the sounds from one another. Problems arise when unwanted sounds leak onto tracks that contain sound sources that were intended to be isolated from all other sound sources. This leakage can be the cause of many problems later on in the recording process.

Sound quality should be carefully evaluated during the tracking stage. The amount of timbre detail, as well as many aspects of the precise sound quality, of the sound sources will be determined at this stage. These are major decisions that will have decisive impact on the sound of the final recording, and are both largely determined by microphone selection and placement.

Many beginning recordists rely on equalization, to obtain a suitable sound quality during tracking. Such use of equalization actually distorts the natural sound quality of the sound source, and does not alter the timbre equally well throughout the instrument's range; it should be used only if the equalized sound is desired over the sound of the naturally sounding source as captured without distortion, or when practical considerations limit microphone selection and placement options. Equalization is often used in an attempt to obtain the desired sound quality by compensating for an incorrect microphone selection or placement.

Related to performance technique, the ways that performers produce sound on their instrument, or the instruments themselves, can create unwanted sound qualities that must be negated during the tracking process. Often, these aspects of sound quality are very subtle and go unnoticed until the mixing stage (when it is too late to correct them).

Instruments are capable of making unwanted sounds, as well as musical ones. The sound of a guitarist's hand moving on the fingerboard, the breath sounds of a vocalist or wind player, or the release of a keyboard pedal are but a few of the possible nonmusical and (normally) unwanted noises that may be produced by instruments during the tracking process.

These sounds are easily eliminated during the tracking process through altering microphone placement, slightly modifying performance technique, or making minor repairs to the instrument. The multitrack tape should be as free of all unwanted live-performer sounds and sound alterations as possible; these sounds will be much more difficult to remove later on in the recording process. These sounds may be comprised of certain performance peculiarities that (depending on the situation) can be alleviated only by signal processing (such as the use of a de-esser on a vocalist).

The recordist will be anticipating the mixdown process while compiling the basic tracks. The objective of the recordist will be to have the opportunity to be in complete control of combining sounds during the mixing process.

Any mixing of microphones during the tracking process will greatly diminish the amount of independent control the recordist will have over the individual sound sources during the final mixdown process. Mixing will occur during the tracking stage, as submixes, to consolidate instruments and open tracks.

Submixes will be carefully planned at the beginning of the session, with a clear idea of how the sounds will be present in the final mix. Drums are often condensed into submixes (mixed either live or through overdubbing and bouncing). Other mixes that will occur during the tracking process include combining several microphones (and/or a direct box) on the same instrument(s). The recordist must be looking ahead to how these sounds will appear in the anticipated final mix.

Preprocessing is an alteration to the components of the sound source before the source has become available in the routing and mixing stages of the recording chain. Preprocessing will also diminish the amount of control the recordist will have over the sound, during the mixing process. At times, it is desirable to preprocess signals, but very often it is not.

Desirable preprocessing might include stereo microphone techniques used on sound sources, effects that are integral parts of the sound quality of an instrument (distorted guitar), processors that are used to provide a specific sound quality (compressed bass), or processing that is used to eliminate unwanted sounds during tracking (noise gated drums). Once a source has been preprocessed, the alterations to the sound source cannot be undone. The recordist should be confident that they want the processed sound before committing it to tape; if flexibility allows, split the signal and record both a processed and an unprocessed (dry) version of the sound, allowing for the option of undoing the processing. A substantial amount of time is often spent tuning processing equipment. If this tuning can take place without the musicians present and performing, the session will benefit.

Some initial planning of the mixdown sessions will begin during the tracking process. Certain events that will need to take place during the mixdown session will become apparent as the tracking process unfolds. Keeping a tally of these observations will save considerable time later on, and may help later tracking decisions.

Examples of items that should be noted, for the mixing process, are:

- Sudden changes in the mix that may be required because of the content of the tracks;
- Certain processing techniques that are anticipated;

- Any spatial relationships or environmental characteristics that may be anticipated as sound qualities of certain tracks; and
- Track noises or poor performances of certain sections that will need to be eliminated during the mix.

These are just a few examples of the many factors that may become apparent during the tracking process.

SIGNAL PROCESSING: SHIFTING PERSPECTIVE TO RESHAPE SOUNDS AND MUSIC

Sound sources are shaped by the recordist with signal processing. Signal processing is applied to the sound source, to complete the process of carefully crafting sound quality. The sound quality of the sound sources should align with the functions and meanings of the musical materials and creative ideas, after signal processing. Signal processing is applied as the final shaping of the sound sources, to suit their relationships with the other sounds in the context of the project and to suit the message of the music. In the recording chain, signal processing can occur in a number of locations. It may be incorporated in the tracking as preprocessing, and is most often used to bridge the tracking and mixdown processes.

Signal processing often occurs in separate studio sessions, between the tracking sessions and in preparation for the mixdown session(s), without performing many actual rerecordings of the basic tracks. Rather, tracks are evaluated and signal processors are applied to the tracks; the resulting processor settings and rough mixes are noted in session documentation and not usually dedicated to tape (with the exception of any possible submixes).

The final processing of the sound sources interacts with the mixdown process itself. The processor settings that were determined during the sessions following the laying of basic tracks are reproduced during the mixdown session(s). In effect, the recordist performs the signal processing of the sound sources in real time during mixdown. The potential exists to change processor settings in real time. Although this is not often utilized, it is common to alter the ratio of processed signal to unprocessed signal, during the course of the mix.

In the performance of signal processing, the recordist will focus on the component parts of the sound qualities of the sound sources. Small, precise changes in sound quality are possible with signal processing, requiring the recordist to listen at the lowest levels of perspective and to continually shift focus between the various artistic elements (or perceived parameters) being altered.

Most signal processing involves critical listening. The sound source is considered for its timbral qualities out of context, and as a separate entity. In this way, the sound can be shaped to the precise sound qualities desired by the recordist, without the distractions of context.

The sound's timbre is conceived through translating the perceived parameters of sound into the correlated physical dimensions of sound. The recordist will know what is happening physically to the sound, if he or she has a prior knowledge of the basic acoustics. This knowledge of the physical dimensions of the sound, as compiled through perception, is vital for successful signal processing.

Signal processing alters the electronic (analog or digital) representation of the sound source. In this state, the sound source exists in its physical dimensions. The various signal processors are designed to perform specific alterations to these physical dimensions, which will all contribute to changes in the timbre of the sound source. Signal processors are only useful as creative tools if the recordist is in control of these changes in the physical dimensions. Immediately following the recordist's alterations of the sound source while conceiving their physical dimensions, he or she will shift focus to place the sound in the context of an artistic element.

After the sound has been reshaped, the listener will use analytical listening to evaluate the sound. The altered characteristics of the sound source and the overall sound quality of the source will be evaluated as they relate to the other sound sources and to their function in the musical context. The changes that were conceived through the physical dimensions of sound, above, will be evaluated as the activities of the artistic elements contained in the sound source. The sound quality of the sound source may then be evaluated according to its appropriateness to the musical idea.

Signal processing defines sound quality through shaping the physical dimensions of sound. All signal processing will result in timbre changes (perceived as sound quality) or as altered spatial properties. Changes in timbre will play a central role in nearly all evaluations.

Timbre, as the overall quality of a sound, is a composite of a multitude of functions of frequency and amplitude displacements over time. Timbral quality will reflect most changes of frequency, amplitude, and time that will occur during signal processing.

Spatial properties of sounds are related to timbre evaluation and will be addressed during signal processing. Distance cues rely heavily on timbre detail. Environmental characteristics are largely determined by timbre differences between the listener's knowledge of the unaltered sound source and the listener's perception of the sound in the environment. The time characteristics created through signal processing will contribute directly to defining distance and environmental characteristics.

Stereo location is created largely in the mixing process, as time and amplitude differences between the two loudspeakers and/or the listener's ears. It is the only element of sound that is not largely interrelated with the others.

Three types of signal processing exist. Each signal processor is designed to function on a particular physical dimension of sound. An alteration in one of the physical dimensions of sound will cause a change in the other dimensions; further, alterations of the physical dimensions will cause changes in timbre (sound quality). The three types of processors do not only cause audible changes in the physical dimensions for which they are named, they may also alter the timbre of the sound source:

- Frequency processors;
- Amplitude processors; and
- Time processors.

Frequency processors include equalizers and filters. Compressors, limiters, expanders, noise gates, and de-essers are the primary amplitude processors. Time processors are primarily delay and reverberation units. Effects devices are hybrids of one of these three primary categories. Some examples of these specialized signal processors are flange and chorusing devices, being time processors, and The Aphex Aural Exciter, being (essentially) a frequency processor.

THE MIX: PERFORMING THE RECORDING

The mixdown of the multitrack tape is actually a live performance of the recording. The sounds that were stored on the multitrack tape are fed to the mixing console, perhaps signal processed for sound quality, assigned spatial properties, and combined with the other sound sources. A mix of all of the sound sources is determined by utilizing the artistic elements, which will be discussed later. The mix will then be performed in real time, after considerable preparation and rehearsal.

The mixdown session will result in a two-channel version of the work, which will become the master of the recording. The recordist will create the desired final recording (mix) by the following ways of shaping the artistic elements:

- Combining the sound sources with a focus on sound quality;
- Performing the individual dynamic levels of the sound sources; and
- Providing each sound source with suitable spatial properties.

Mixing is a process that combines (1) creating an artistic blend of timbres and dynamic levels, and assigned spatial location, distance, and environment qualities (using the skills and concepts of a traditional composer or orchestrator), (2) rehearsing and coordinating the precise changes that will occur in the above three artistic elements (functioning similarly to a traditional conductor) and (3) actually performing the changes that were determined and rehearsed above (similar to a performance on a traditional musical instrument).

The individual sound qualities of all the sound sources are combined in the mix. The recordist will be focused on the dimensions of the individual sound qualities, making any alterations, to assist in the dramatic qualities of the sources and to assist in making more pleasing combinations of timbres. As the sound qualities of the sources are combined to create the higher-level sound qualities of ensembles (groups of instruments) and the overall sound quality of the program (the piece of music as a whole), the focus of the recordist will shift perspective between these various levels, while continuing to scan between the components of timbre.

Throughout the process of compiling (composing) the mix, the recordist's attention will return to timbre and sound quality, listening to the timbres as separate entities (out of time), and listening to sound quality in the musical contexts of all hierarchical levels. The functions of timbre and sound quality that will be of concern are:

- Final shaping of the sound qualities of each source, to define their unique character;
- Timbre changes of sources to suit the requirements of the mix (blend of instruments into ensembles or register areas); and
- Registral placement of the sound source in terms of textural density.

The sound qualities of the individual sources may be shaped in the mix. It is common for final signal processing to occur during the generation of signal between the multitrack tape and the two-channel mix. Signal processing may take the forms of time, frequency, or amplitude processing. Any processor may be used effectively if it is appropriate for the context of the piece and for the sound source. Processors may be applied to the sound source in chains (or in series of processors), as well as individually, allowing for a very detailed tuning of sound quality.

The function of signal processing will be to finalize the shaping of the sound qualities of the individual sound source. This sound quality that characterizes the sound source will combine with the musical material it is performing, to create an overall musical effect. Thereby, the sound quality of the sound source plays a significant role in successfully presenting the musical idea.

A source's sound qualities may remain constant throughout the piece, or the qualities may make sudden, static changes or gradual changes over time in the mix. A multitude of possibilities exist. The actual signal processing for any project will be determined by the needs of the musical material and the resources of the recording studio.

In the mixdown process, a source's sound qualities are shaped in relation to the other, directly related sound sources, as well as within the musical context. Often achieved through subtle changes in equalization, processing may assist the recordist in achieving a sense of ensemble blend from a group of individual sound sources. In an opposite process, sounds may be processed with the intent of making them more distinguishable from the other sources in the musical texture. Although equalization is often used to add emphasis to particular sound sources, time and amplitude processing, as well as hybrid frequency processors, can be used successfully and creatively to achieve these same goals.

The recordist will consider *textural density* when mixing sound qualities. Textural density is the placement of pitch/frequency information throughout the hearing range, and the amount of pitch information in specific pitch areas. The dimensions of textural density are the width of the pitch information present in the musical texture and any areas that may be emphasized by a high concentration of pitch/frequency information. This aspect of pitch relationships will contribute directly to the character and momentum of the piece of music.

A piece of music may have a high concentration of musical material (and/or sound qualities) in specific pitch areas; (for example, a piece may have a high concentration of pitch material in the low pitch area). Certain works will alternate between a number of emphasized pitch areas (those with a high concentration of musical material).

Textural density may be utilized in innumerable ways to assist in shaping the overall shape of the music, as well as in defining the relationships of the individual sound sources. This topic will be covered in more detail in Part 3.

Sound sources are assigned dynamic levels in relation to one another, in the mix. The entire mixdown process is often misconceived as the process of determining the dynamic level relationships of the sound sources. As noted, the mixdown process actually combines many complex sound relationships, of which dynamics is only one.

A *musical balance* of the music recording is determined; this is the relationship of the dynamic levels of each instrument to one another and to the overall musical texture. The individual sound sources are combined into a single musical texture, each source at its own dynamic level. The mixing console allows the recordist to perform the individual dynamic levels of the sound sources and to make any changes in level in real time.

A difference between the actual perceived loudness of the sound source and the perceived intensity at which the musical material was performed may become evident at this stage of the production process. This difference between musical balance and performance intensity (covered in-depth in Part 3) may be used to great creative advantage. The mixing process itself can take on dramatic and creative dimensions in altering the realities of sound quality and dynamic level relationships.

The spatial properties of the recording are created during mixdown. These properties will provide each sound source with unique characteristics, and will give the entire recording a sense of two-dimensional space.

The spatial properties created in the mix provide an illusion of a space within which the performance takes place. Sounds are placed at locations within this perceived space in natural and unnatural stereo location and distance relationships. Sounds are perceived as existing within their own environments (spaces). The sound characteristics of their environments may sound as if they could occur in nature, may sound manufactured, or may sound as if the laws of physics have been violated—as if the acoustical environment could not exist on Earth.

The recordist creates space relationships for the "reality" of the recording. An illusion of a space is created; the performance (recording) will be perceived as taking place within this space. The dimensions of this space are defined by the recordist in the mix, and the perceived qualities of these dimensions are restricted solely by the craft and imagination of the recordist. At the extreme, spatial properties may create the illusion of a new acoustic reality, through the relationships of and the characteristics applied to the sound sources.

Spatial properties may be related in a hierarchical structure. Each sound source will be placed at a position in the stereo array, at a distance from the listener and within a perceived environment. These sound sources may be grouped together into ensembles—within areas on the horizontal plane, within the same environment, or located at similar distances from the listener. The individual sources and groups of sources are related to all other sources, by their location in the sound stage. They are grouped into one large location that has width and depth, and that is contained within the overall environment that the listener perceives him or herself to be occupying.

Spatial properties contribute to the recording in a number of ways:

- Applied to each source to further define its character;
- Applied to each source to suit the requirements of the mix (blend instruments into ensembles or make sources more prominent in the musical texture);
- Provide illusions of space for the sound source (or groups of sources); and
- Provide a set of spatial relationships for the entire program.

The mixdown process will place each sound in space. Each sound source will be at a specific location in the stereo array created by the two loudspeakers, will be at a perceived distance from the listener, and will be perceived as performing within a particular environment. The recordist must determine these locations and characteristics as part of the mixdown process and the signal processing sessions that precede it.

These spatial properties may be used to delineate the sound sources into having their own unique characteristics, or they may be used to cause a group of sound sources to blend into a sense of an ensemble. It is possible for a group of instruments to be grouped in one of the three dimensions, but to have very different characteristics in others. For example, all sources may be located in a specific area on the left side of the stereo array, but at very different distances and having distinctly different environmental characteristics.

The recordist will also determine the spatial properties of the overall program. The individual sources and any groups of sources will be perceived as being located within a single area (the sound stage), within a single performance space (the perceived performance environment). The dimensions of the sound stage will be created by stereo location and distance placements of the sound sources.

The dimensions of the perceived performance environment may be applied during the mastering process (with planning occurring during the mixdown), but will most often be the result of the listener's perception of a composite environment created by elements of the predominant environmental characteristics of the primary sound sources of the work. In this way, shaping the environmental characteristics of the sound sources that present the most important musical materials will have a direct and marked impact on the listener's impression of the performance environment, within which the recording itself appears to take place. This concept is presented in greater depth in Part 3.

The illusion of the stereo location of sound sources is created through time and/or amplitude differences between the two ears. The same signal appears at each ear at a different amplitude and/or at a different point in its time cycle. These differences are created during mixdown, to be present at each loudspeaker. Through these differences, the recordist is able to place the sound anywhere within or slightly beyond the locations of the loudspeakers, and to create phantom images of widely varying widths. Locations and widths are created by panning and/or combining signals and channels.

The stability of these phantom images is a measure of the degree of reliability at which they can be reproduced. Aside from being dependent upon the listener's location (equidistant from the two loudspeakers), the phantom images can be distorted (or unsuccessfully created) by phase shifting and comb filtering effects between the two loudspeakers. These phase and amplitude anomalies are perceived by a concentrated focus on stereo location, noting any changes in width and location, or lack of defined location. Time processing is often helpful in addressing these problems. Delaying one of the signals (sent to one of the loudspeakers) by a suitable time increment, by carefully tuning a delay unit, may stabilize the image.

Distance is the other dimension of sound stage imaging. It is primarily perceived through the amount of definition of the small details of the sound source's timbre. It is also dependent upon the time delay between the arrival of the direct sound and the reverberant energy, and by the ratio of the amounts of direct sound versus reverberant sound.

In the mixdown process, distance cues are created by applying time processing and environmental characteristics, in balance with the amount of detail present in the initial tracking of the sound source (often, the amount of detail heard is partially determined by the dynamic level at which the sound is added into the mix). The actual distance cues are determined by the amount of perceived timbre detail. This detail is influenced by the amount of time/environmental information added to the sound source, and by the amount (and qualities) of the original sound source added to the mix. Remember, distance is *not* dynamic level.

The recordist will add environmental characteristics to sound sources through the use of reverberation units and delays. The sound sources will be defined as individual ideas by their unique environment placements. These individual environments will then be perceived as existing within the environment of the overall recording. This space within space perception allows the recordist to apply environmental characteristics creatively.

Natural and unnatural relationships of environments, stereo location, and distance locations may be exploited as artistic elements. These ideas may be used to support the musical materials, or they may be used to add another unique dimension to a source's sound quality. These elements are determined entirely by the recordist. Decisions concerning the qualities of these elements are made primarily in the mixdown stage. These qualities may be applied to the piece of music in any way the recordist is able to imagine.

The mix is a performance of the recording. Performances must be adequately rehearsed to be accurate, artistically expressive, and interesting. During the mix, many things happen. The three artistic elements that are determined in the mix will be the musical material that is performed during the mixing process. These three artistic elements of sound qualities, dynamic levels, and spatial properties may be altered in real time, or may be at predetermined levels. The recordist will thoroughly plan the relationships of all sounds throughout the recording.

The recordist will prepare for mixdown by composing the mix. In doing so, he or she will shape the artistic elements:

- Specific dynamic levels determined for each sound source (or each group of sound sources located on the same track);
- Specific sound qualities determined for each sound source;
- Specific environmental characteristics determined for each sound source;
- Specific stereo locations determined for each sound source; and
- Specific distance locations determined for each sound source.

These levels may remain constant throughout the piece. It is quite common for certain artistic elements to remain constant throughout a mix. It is also common for changes to occur in specific elements, of certain sound sources. These changes may take place in the sound sources presenting the secondary musical materials, as well as in those sources presenting the primary musical ideas. The types of changes, amounts of changes, timing of changes, and so forth, of the artistic elements will directly shape the piece of music.

The recordist will plan out all these changes and rehearse them during dry runs of the mixdown process. The changes in the mix will be related to two basic ideas:

- Changes between two or more specific, pre-determined levels of any of the three artistic elements, occurring at specific points in the piece; and
- Changes between two or more specific, pre-determined levels of any of the three artistic elements, occurring at a specific rate of speed and between specific points in the piece.

The recordist will control the locations of changes that need to take place (in relation to the music), the beginning and ending levels of the changes in the mix, and the speed of the changes between levels.

Much of the planning of the mixdown will have taken place during the sessions that determined signal processing. Rough mixes were compiled at that stage and during the tracking and overdubbing sessions. Rough mixes used during those sessions will most often emphasize vocal parts, drum, and rhythm section tracks; at times they will contain reference tracks (specifically laid to help overdubbing and not intended to be part of the final recording). These rough mixes give the performers reference tracks that are easy to follow and that provide enough musical information to allow the performer to get the impression that they are performing the specific piece of music.

The rough mixes and final mix are not necessarily the same or even similar. Rough mixes are made for many purposes, from tuning signal processors to laying tracks. A rough mix may be far from the most pleasing representation of the piece of music.

The multitrack tape should have all unwanted tracks and unwanted sounds removed before mixdown. All other problem areas of the multitrack tape should be identified and located; this will complete the planning of the mix. Within the choreography of the mixdown, plans must be made to mute tracks when they contain unwanted sounds or when they are not being used. Some signal processing (especially noise gates) may need to be dedicated to cleaning up tracks that have unwanted sounds.

The recordist will often need assistance in performing a complicated mix. With a little practice, several people can be involved in the process of changing fader levels, muting channels, altering processor settings, rerouting signal paths, and so on. Successfully executing complex mixes often requires help from qualified assistants. With planning the mix and rehearsing all changes of levels, and so forth, a group of people can easily function as one—as long as the person leading the mix has a clear idea of what is to happen when.

Mixes are often created for individual sections of a piece. This may be required if the recordist is short on assistants, in the case of complex mixes, or if large-scale changes in signal processor routing and assignments, or many changes in fader levels, occur in the mixdown.

The beginning and ending points of the mixes will be planned for ease of editing and for musicality. The different mixes will move one-to-another smoothly and unnoticeably. Large-scale changes in the mix are striking and quite dramatic; they occur most frequently at major structural divisions of the piece, such as between verses and choruses.

It is imperative that all submixes be at similar loudness levels. The mixes of the various sections will be compiled in making the master tape. Sudden changes in level will be very noticeable. They will detract from the musicality of the piece and may cause the tape to be unusable for transfer to other formats (lp, cassette, compact disc).

An automation system can be of great assistance to the recordist in the mixdown process. Most importantly, an automation system can allow the recordist to refine dynamic level relationships of the tracks (sound sources) and can free up the recordist to perform other tasks during the mixdown process. An automation system may perform some of the changes to the artistic elements that were described earlier.

The applications of automation are largely dependent upon the sophistication of the individual system. Automation systems typically are able to accurately change dynamic levels of channels and will mute channels, with consistency, in real time. Some systems are able to change equalization through the automation, or to alter the panning of the channel, and some systems are able to route signals.

The mixdown is the performance of the recording. Preparation for the mix begins with selecting suitable sound sources for the musical materials, as well as the appropriate microphones to capture the sound sources. The sound sources are then recorded onto multitrack tape, with their sound qualities carefully shaped, in a planned sequence of events. The sound sources are considered for sound quality, dynamic level, and spatial properties, in making the detailed plans for performing the mixdown. The mix results in a two-channel version of the piece, which will become the master of the recording.

7

The Final Artistic Processes and An Overview of Music Production Sequences

Multitrack and direct-to two-track recording processes involve very different aesthetics and are very different in their production sequences and their utilization of recording techniques. The artistic resources of the recording medium, the importance of the performers and their interactions, and the amount of input the recordist has on the artistic aspects of the recording may be very different between multitrack recordings and direct-to two-track recordings.

The differences in approach to production (often linked to aesthetic approaches discussed in Chapter 4) are reflected in the sequence of events that occur in creating the music recordings. They are directly related to the amount of control the recordist will have over the musical relationships in the recording and the extent to which the musicians determine the musical relationships (by their performances as individuals and by their interactions as an ensemble).

The final, creative processes of editing and mastering the music recording will be different between individual approaches to multitrack recordings; those differences will be even greater between multitrack recordings and the various approaches to direct-to two-track recordings. Editing and mastering will be considered conceptually, in relation to the final recording and the final presentation of the music and its artistic elements.

AN OVERVIEW OF TWO SEQUENCES FOR CREATING A MUSIC RECORDING

The recording-and-reproduction signal chain may take many forms, depending on the complexity of the recording project. The flow of the signal through the chain and the order of equipment and events will normally be consistent with Figure 7-1.

The activities and associated devices of the recording-and-reproduction signal chain generally appear in the order of the following bulleted list. Some devices are utilized throughout the recording process, such as the mixing console or the monitoring system. Certain activities, such as editing, may occur at several different stages of the production process and within signal chain, or it may not occur at all.

- Microphones (or electronic instruments);
- Console (preprocessing, record levels, and routing);
- Digital multitrack recorder or analog (with any noise reduction processors) multitrack recorder
- Console (routing and mixing);
- Signal processors;
- System control methods (automation systems, computers and MIDI, SMPTE locking for synchronization);
- Master recorder (analog or digital);
- Editing (razor blade or computer-based); and
- Monitoring.

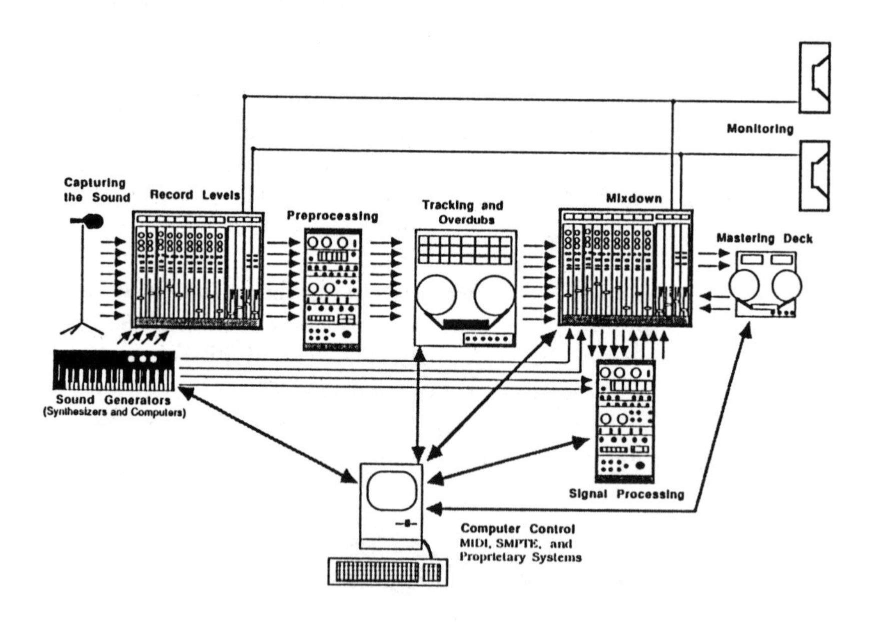

FIGURE 7-1. Recording-and-Reproduction Signal Chain.

Every music recording project is unique. The sequence of events in the individual recording production and the utilization of the recording chain will be adapted to suit the needs of the individual music recording.

Some projects will require more session preplanning than others; one project may require more mixdown preparation and rehearsal than another; other projects will have still different requirements. The order of events will be consistent with the following list. Some overlapping of activity between the events and some alterations to the orders of portions of the events will commonly occur.

A complete sequence of events for a multitrack recording might be:

1. Session preplanning–determine music to be recorded (writing the music), rehearse musicians, define sound sources, select microphones, plan track assignments, determine the recording order of the tracks;
2. Tracking session–record reference tracks (vocals and accompaniment), followed by recording the basic tracks (primarily the rhythm tracks);
3. Editing of basic tracks for outtakes and to create the basic structure and length of the piece, reorganize tracks;
4. Overdub sessions–adding solo parts and secondary ideas to the basic tracks, refining the musical material, composing and recording any additional parts to fill newly discovered requirements of the piece;
5. Processing- and mixdown-preparation sessions–finalize the sound qualities of sound sources, edit the source tape for mixdown (reorganize tracks, remove unwanted sounds);
6. Mixdown rehearsal sessions–"compose" the mix by defining the artistic elements of dynamic levels, spatial properties, and sound quality for each sound source, and by considering the interrelationships of the mix and the musical materials of the piece, rehearse the mixdown sequence with people who will assist in the session;
7. Mixdown session–perform the mix(es), mixing the multitrack tape down to two tracks (often occurring during the same session as no. 6.);
8. Mastering session–assemble and process a master tape by combining the section mixes, and by applying any global signal processing.

Direct-to two-track recordings (or direct-to-mono) are common in music recordings for film, television, and advertising. The applications of this approach to recording are not limited to recording art music (such as orchestral, choral, or chamber music). This approach may be suitable to jazz, folk, ethnic, popular, rock, or any other music where the musicians (or the conductor) want to be in control of nearly all musical relationships within their performance, or where the function of the recording is best served by having all the musical parts performed at once (often the case for film scoring or archival recordings, as examples).

The process of making direct-to two-track recordings is strikingly different from the multitrack recording process discussed above. Nearly all the recording process considerations of Chapter 6 are not directly relevant to this approach.

Further, the perspective of defining the sound quality, as presented in Chapter 5, is shifted from the sound source to the perspectives of the overall ensemble, of groups of instruments within the ensemble, or to the perspective of a limited number of individual soloists.

A complete sequence of events for a direct-to two-track recording might follow the following outline. These events will be discussed in detail and in the form of a commonly occurring sequence, in the following paragraphs.

- Session preplanning;
- Creating the sound quality of the recording;
- Consultations with the conductor (musicians);
- Recording session;
- Selection of takes; and
- Editing to compile a master tape.

1. Session preplanning always begins the production sequence. Once the music to be recorded is known, the recordist will need to know the performance level of the musicians (performers) and the location of the recording session (if it will not take place in the studio). This will allow the recordist to select suitable microphones, project an appropriate stereo microphone technique, and plan microphone placements (of the accent microphones and the stereo array or arrays). The acoustics of the recording environment will need to be evaluated, if the recording is to take place in a space unknown to the recordist.

   The recordist will meet with the performers (or the conductor of the ensemble) to determine how the work can be effectively (and transparently) divided into sections. The problems of editing the sections together into the master tape, and of stopping and starting the ensemble, must be considered when the work is divided into sections. The order in which the sections will be recorded may then be determined. The recordist's and conductor's scores, and the musicians' parts, will be marked to identify these sections, making starting and stopping the ensemble during the recording session more efficient. This conference between the recordist and the conductor (or performers) will also clarify the role of the recordist in the project (who may be responsible for some of the artistic decisions that must be made).

Further discussion should clearly define the sound qualities that are desired for the recording project.

2. An actual sound quality of the recording is obtained through monitoring a dress rehearsal of the ensemble, in the performance space in which the recording will take place. Alterations to microphone selection and placement will be made, to achieve the desired sound quality of the recording, as it was determined during the session planning meeting. The microphone selection and placement will largely determine the spatial properties, dynamic level relationships, and sound quality of the recording; balancing of microphones and signal processing will do the final shaping of the sound quality of the recording. Any necessary signal processing (environmental characteristics, time delay, dynamic processing being most common) for accent microphones and the stereo array (or arrays) will be added and tuned at this stage of the production sequence.

   After the final sound quality has been established, portions of the rehearsal are recorded for later reference. Any changes in the mix that may be required for the recording session are determined. These changes will be thoroughly rehearsed during this rehearsal of the ensemble.
3. The recordist and conductor (or musicians) will listen to the reference tape that was made during the dress rehearsal. Often, their discussion will be solely on the subject of sound quality; all musical considerations may be determined between the conductor and the musicians, or among the musicians themselves. Any alterations that must be made to the sound quality of the recording are determined during this conference. The recordist must obtain a clear idea of the sound qualities required for the project.
4. All requested changes to the sound quality of the recording are evaluated and planned by the recordist. The microphone and recording equipment setup for the recording session will reflect these changes. Sound quality is rechecked during the musicians' warm-up period, before the beginning of the recording session. The recordist and conductor (musicians) make a final evaluation of the changes that were made to the sound quality, and confirm that the sound quality is correct. The recordist, conductor, and musicians will briefly clarify the logistics of the recording session (how stopping, starting, slating, and so on, will be handled).
5. The recording session follows. During the recording session, the sections of the piece are performed in the pre-arranged order; many takes may be performed of each section of the work, until two suitable takes (of each section) are recorded. Each take of each section is monitored by

the recordist, with a focus on consistency of loudness levels, tempo, intonation, performance quality, and the expressive qualities of the performance. An assistant engineer or second engineer may be used to assist in sound evaluation. This person would focus their attention on the technical and critical listening aspects of the sound of the recording. If possible, another assistant might maintain a record of the content of the session tapes, making notes on the recordist's observations of each take, and of the observations of the musicians' spokesperson (usually the conductor, if one is present).

Any changes in the mix that are required in the recording were choreographed during the rehearsal session(s). These changes in the mix are performed, in real time, during the musician's performance in the recording session. The recordist may coordinate the activities of one or more assistant engineers, who would physically perform the actual changes in the mix. The recordist would remain focused on the accuracy level of all these changes, as they are made, as well as their relationships to the performance.

A multitrack recording of the session may be made simultaneously with the two-track mix, to make a safety recording of the session. This will allow the recordist to perform a remix of the session at some future time, should this be necessary. All microphones will be sent from the console directly to the multitrack recorder, usually routed from the console's direct or patch outputs. No balancing of dynamic levels or extra signal processing will be performed on these tracks—this would defeat the purpose of recording the multitrack tape.

6. The conductor (musicians) will listen to the session tapes with the recordist. Often, they will listen to only a few takes of each section of the piece; the takes will have been preselected by the recordist, using the conductor's observations (which were written down by the production assistant) during the recording session as a guide for selecting takes. The takes that will be used in the final recording are determined during this conference between the recordist and the conductor. Both parties will discuss their perception of the sound qualities of the takes, and the recordist may or may not be asked to evaluate the musicality or accuracy of the performances of each take. Any specific aspects of the sound quality that are undesirable will be identified. The recordist will determine signal processing alterations to attempt to solve (or minimize) the sound quality problems, and may play some of the possible alterations for the conductor (performers) during this session. A remix from the safety multitrack tape may be determined at this point, as a last resort.
7. Any signal processing or mixing alterations that were determined while reviewing takes are performed by the recordist in the mastering session.

These changes may be performed before or after the master tape has been compiled, depending on the alterations that need to be made. The master tape is created by splicing together all the selected takes, in the correct order. Any global signal processing will be applied to the overall program after the edited tape has been assembled, and a master tape will be made by playing the spliced session tape through any signal processing device(s), to another mastering deck, which will record the actual session master. The recordist will arrange for the conductor (musicians) to hear the master tape, for final observations and approval. Any final alterations to sound quality, and so forth, requested by the conductor, will be performed by the recordist, and will complete the project.

EDITING: SUSPENSION OF TIME

The recordist may physically hold time in his or her hands. Audio recording transfers sound, which can only occur in time, into a storage medium where the sound is physically located, suspended out of time. The sound may be altered by physically altering either the storage medium itself or the way in which the storage medium reproduces the sound. The sound can be replayed or altered at any time, without relation to real time. It may be altered at any point in the future and may be replayed forwards, backwards, and at any speed (even at uneven speeds).

Sounds may be edited in their physical state. In *editing* sound, the recordist is able to precisely shape material out of real time. The editing process is the joining of two time segments. Each of the time segments is comprised of a group of sounds. The time segments may exist as pieces of tape or as computer information.

In joining the sound segments, the recordist is significantly altering the piece of music and its artistic message. These alterations to the music that are made possible by editing serve many functions, as will be discussed in the following paragraphs. The edit must be accomplished in an artistically sensitive manner and must be inaudible in all areas of technical quality.

It is impossible to perform technically inaudible or artistically sensitive edits in the music under many circumstances, and in many locations. The recordist will search for potential edit points and will carefully calculate, and possibly rehearse, an edit before actually performing the edit.

The editing function most readily apparent is that of compiling a master tape of a recording session. In this process, many sections of the piece of music are joined into a single performance. This allows the most appropriate material, or the most pleasing performances, to be included in the recording (which may represent a perfect performance of the piece). A single performance is compiled from the many takes of many small segments of a piece of a direct-to two-track recording, or from joining the few, individual mixes of multitrack recording.

It is possible to reorder sounds through editing techniques. The major sections of a piece of music may be rearranged. Entire measures may be exchanged, or sounds within a measure may be reordered.

The editing process will alter all sounds present. Sounds cannot be reordered unless they are isolated. It will not be possible to reorder the sounds of instruments in a drum fill, without also moving the sounds that occur simultaneously with the drum sounds. Likewise, it is impossible to cut a sound source into numerous time segments and reorder the sound, unless only that sound exists on the tape or in the memory buffer.

When a two-track recording is edited, all sounds in the recording are edited. An edit on a multitrack (analog) tape cuts across all tracks.

While this may make editing inappropriate for certain applications, editing is the process used to compile source tapes for mastering. Multitrack tapes are also edited, in preparation for overdubbing, tracking, or mixdown.

Unwanted sounds and noises may also be removed from a recording. This function of editing is as important as removing or rearranging the sections of a performance, and is found in all of the above editing applications. When a sound is removed, it may be necessary to replace it with another sound or a silence of equal lengths; if this is not done, the passage of time (rhythm) will be altered. To fill a space between two edit points created by removing an unwanted sound, use some prerecorded, ambient sound of the room (sometimes called room tone) or some recorded blank tape (in analog). This will be accomplished before adding any global processing.

Edit points are calculated by anticipating the sound that will be created when the two segments are joined. Each segment to be joined will be evaluated for its sound qualities, to determine the location of the splice. The edit points in the music are often found through trial and error–listening carefully to the sound qualities of the two segments, remembering what was heard, and comparing the two sound events. Edits are most easily made at points where loud attacks are performed by prominent instruments or the entire ensemble, or immediately before or after (not during) areas of silence.

Sound sources that are sustaining over the edit point or that are present in each time segment make the edits more difficult. Changes in the sound source will make the edit point audible. Audible edits may be created by many factors.

Both the critical qualities of timbre and the artistic elements in relation to the musical materials must be considered in determining suitable edit points. The perceived parameters of sound must be evaluated for any changes (distortions of the waveform) that were caused by the edit process itself and for any noises that may have been added.

In calculating the edit, the recordist will scan all artistic elements or perceived parameters of sound, at all perspectives, to determine a usable edit point.

It is not possible to perform an inaudible edit when large differences exist in any of the elements of sound, between the two time segments. Such a splice would result in a sudden alteration of the component of sound at the point where the two segments meet; the sudden change would draw the listener's attention and would be distracting to the music. As soon as an edit has been made, it should be checked for accuracy and to be certain it is inaudible and that no noises where added in making the edit.

Sudden changes between segments may be desired, as in creating a master tape where the splice actually joins very different musical ideas. In these instances, the recordist must make certain that the editing process does not create noise at the edit point and that the sudden changes are presented as a part of the musical materials (have significance and are handled artistically).

Among the most common of inconsistencies that are present between two time segments are differences in loudness levels. Even subtle changes can be quite audible. Calculating the loudness levels between various takes of an entire ensemble can be quite difficult. Beginning recordists will often only notice problems in this area after the edit has been made.

Tape noise is part of analog recording. The amount of noise on the tape may or may not be consistent throughout a recording. Changes in the noise floor at edit points will be very noticeable.

Differences in the sound quality of individual instruments and of the overall ensemble are easily overlooked. The potential exists for sound sources and an ensemble to undergo significant changes in sound quality, from the beginning of a recording session to the end. These changes in sound quality may be caused by performer fatigue, performance intensity, artistic expression, or a change in temperature or humidity in the performance space. Even subtle changes of sound quality that may not be consciously perceived by the listener can have a marked impact on the musicality of the recording.

Changes in pitch between the two segments are the most noticeable of all changes. The recordist must be well aware of any inconsistencies in this element. Inconsistencies may occur within a particular sound source, or it may be a change of the reference pitch level (tuning) of the ensemble. Care must be taken to monitor the tuning of the ensemble and the intonation of the performers.

No changes in spatial properties should occur at the edit point, unless they are planned. It is common for spatial properties to be considerably different between time segments, when they represent different mixes of a multitrack master. Sudden shifts of distance locations are common, and have the potential to create few technical problems. Although sudden shifts of stereo location are equally common between time segments, technical problems can be created. Among these are phase differences between similar sounds at the edit point and sudden transients engaging a loudspeaker in mid-cycle of a soundwave (which can create distortion).

Sound sources or environments that have a lengthy decay may need to be carried over across the edit point. This may or may not be possible, depending on the musical context and the nature of the sounds themselves. Edits at these points are sometimes possible, but must be handled carefully, to avoid audible changes of sound quality. These edits may need to be planned before the recording session, with suitable alterations made to the performances at the session.

The musical material must remain in rhythm. It is possible to add or subtract time in making an edit. Rhythm changes are very noticeable in their affect on the performance; measures will appear to be extended or shortened by fractions of a beat.

Tempo changes between takes can be very distracting. The tempo(s) of the performances will be carefully monitored during the recording process. Any tempo differences that are present between segments will make the edit point very noticeable. An entire take may be unusable, caused solely by tempo inconsistencies.

Analog and digital recording systems have some different characteristics specifically related to their technology. The inherent qualities of each format create advantages or disadvantages, depending on the application of the recording, and the specific nature of the recording session. Either analog or digital recording, may be the more appropriate choice, depending on the individual recording project. Sound is edited very differently in the two technologies.

In an *analog recording*, a physical image of the sound is present as oriented magnetic particles on tape, and the physical characteristics of the image are directly proportional to the soundwave. In editing an analog tape, the tape itself is physically cut with a razor blade. Two cut ends of magnetic tape are joined at either 45-degree or 90-degree angles, and fixed in place with an adhesive tape. The 90-degree angle is most prone to generating a sound as it passes the playback head.

This physical joint between the two pieces of magnetic tape can be the source of noises and distortion. Splices can be audible. As the splice passes the playback head, drop out of the signal can occur, as well as extra sounds created by any physical changes to the waveform. An inaccurate or poorly fixed splice may cause speed changes of the tape, with a related pitch-shift of the entire program. Further, an inaccurate splice may physically alter the waveform by inadvertently removing a small portion of the sound.

Splice locations are found by slowly moving the tape across the playback head of the recorder. By rocking the tape across the head, the recordist is able to hear the edit point. The edit point is physically located on the tape at the playback head, and the tape is marked, removed from the recorder's tape path, placed in an editing block, and cut.

Once an analog tape has been spliced, it is difficult to redo an edit. Splices are difficult to separate, without causing damage to the magnetic tape (which contains the sound–music). If the recordist is successful in undoing the splice, without damaging the tape, it is difficult to cut thin time segments (pieces of tape) off the end of a magnetic tape (should the original splice be just a bit too far to the left of the desired edit point). It is almost impossible to add a small piece of magnetic tape onto the beginning of a tape segment (should the original splice be a bit too far to the right of the desired edit point).

Difficult edits are sometimes rehearsed. Copies of portions of the session tapes are made, and the copies are edited. The recordist gains confidence, or finds the precise edit points that are usable on the copies of the tape. This allows most errors to be made on tape that will not be used in the final version of the project.

Splices can become physically separated. The adhesive tape may not have been adequately fixed when the splice was made, or the adhesive tape may become weak with age. Splices that pull completely loose are easy to discover, and the simplest to correct. Splices that have separated slightly, leaving only a small space between the two pieces of magnetic tape, may be difficult to see, but will be quite audible.

Digital recording formats are quite different from analog. When sound is digitized, the dimensions of the waveform are transferred into a numerical representation, in binary code. Specialized computers (or specialized software usable on personal computers) are used to edit the waveform. The digitized waveform is altered by rearranging information in the computer's random access memory (RAM).

The primary disadvantage of digital editing is that the sound cannot be held in the recordist's hand. The recordist does not know the physical location of the recording and its component sounds. All editing is accomplished through a computer and must be conceptualized more abstractly than analog editing practices.

The primary advantage of digital editing is that the sound is not physically present in the recordist's hands. The sound exists as computer information and may be acted upon in ways that are not limited by physical realities. The following items are possible, using most digital editing systems:

- Precise edit points may be determined, with a resolution to the millisecond being a common time increment.
- An edit may be heard, changed, reheard, and evaluated by the recordist before it is made permanent.
- Edits can be undone, quickly and easily.
- The edit does not alter the original material, the original recording is not edited, and a copy of the original recording is edited, as a computer file (with no generation loss).
- Dynamic levels of the time segments on either side of the edit may be controlled to match at the edit point.
- Edits may be made by cross fading from one segment to the other, or by suddenly switching from one take to another (called a butt edit).
- Some units allow the signal to be heard as it is moved, by the operator, across the cursor point (simulating the rocking of an analog tape across the playback head).
- Time processing and frequency processing are available on some units, to address specific types of inconsistent sound quality and relationships between the two time segments.
- Special effects, such as looping and reversing sounds.

It is evident that digital formats allow more flexibility in and control over the editing process than is available in analog editing. Digital editing systems also allow for a greater margin of error.

MASTERING: THE FINAL ARTISTIC DECISIONS

The final, master tape of the piece of music is the result of a mastering process. The sequence of the process is:

1. The musical materials and their relationships are brought to a two-channel format (existing as either a source tape or as an electronic signal);
2. The musical material (in two channels) is then shaped to finalize its overall characteristics of the recording (arriving at the mastered stage of the piece of music).
3. Finally, all master tapes of all of the pieces of an album project (or film sound track, etc.) are compiled and shaped for consistency between all selections.

The piece of music in the two-channel format may be a live mix, or it may be a source tape. The source tape, which leads to the master tape, may be created in a number of ways: it may be a two-track mix from a multitrack recorder; it may be a direct-to two-track recording from a live performance; it may be an assembled tape from a number of independently mixed sections of a multitrack production; it might be a compilation of any number of session takes that were mixed direct-to two-track. At this stage, all the musical materials and individual sound sources within the recording are in their final form. No additional alterations will be made to the musical materials and sound sources themselves.

The source tape of the piece will now be shaped, in terms of its overall quality. In doing this, a *master tape* will be created. The overall sound quality of the recording will be the global impressions created by the recording and any sound characteristics that are consistent throughout the recording. The *perceived performance intensity* of the entire piece of music will be one of the global impressions. This is a dimension that is not directly shaped by the mastering process of the individual work. It is a composite of the perceived performance intensities of the sound sources in the piece that present the primary musical materials.

The overall sound quality of the recording is a global impression of the emphasized sound characteristics within the recording. These sound qualities may be the result of (1) the listener's perception of the emphasized sound characteristics of the sound sources that present the primary musical materials, (2) signal processing that is applied to the overall program (source tape or two-channel mix), or (3) a combination of the above. All three of these factors are used in the mastering process, depending on what is appropriate for the context of the project.

Signal processing applied to the overall program, (*global signal processing*), is common. It may take the forms of altering either the recording's dynamic contour, frequency response, or environmental characteristics. A consistent loudness level reference is required in certain contexts; it may be appropriate to expand or compress the dynamic range, to alter the dynamic contour, or to limit certain dynamic level peaks. Subtle applications of equalization, to alter the frequency response of the program, is sometimes used in the mastering process. Environmental characteristics may be applied to the overall program, to apply a perceived performance environment onto the entire program (ensuring clarity of space within space relationships); however, the effects of synthetic reverberation on the final form of the recording can dramatically alter the time cues of all distance and environmental relationships of the recording and can alter the overall sound quality of the recording in terms of frequency response.

Signal processing may be added to the final two-channel form of the piece, to create the impression of a sense of ensemble. In multitrack recordings, it is common to get a performance that sounds like the musicians are not reacting to one another and lack unity. Adding a characteristic to the sound that affects all sounds equally, such as compression, may provide a consistency of sound that the performance lacks.

The mastering process of the piece of music may be accomplished during mixdown. Mastering is most commonly interrelated with the mixdown process, in both multitrack and direct-to two-track recordings. In direct-to two-track recordings, the tape that is created by editing together all of the selected takes will often be the master. The master recording may be created in the same step as the mixdown of the multitrack tape. All global processing may be accomplished between the mix output of the console and the input of the two-track mastering deck, making mastering during mixdown possible under certain circumstances.

With digital technology (and the minimal amount of sound degradation with succeeding generations of signal), it is possible to spend more time on the mastering process itself, without diminishing the technical quality of the recording. The changes made to the overall program, in the mastering process, may be crafted much more carefully, if the mixdown and mastering are handled in separate steps. It is useful to separate the two stages of the music production sequence, when circumstances allow.

Individual works and their master tapes are compiled into a master tape of the recording (album project). A consistency of sound quality must exist throughout a project, and the individual works must be related properly to one another. An entire album project will often be related by an overall artistic concept, and it will often have an overall sound, comprised of characteristic sound qualities found throughout the album.

This overall sound, in an artistic sense, is the product of the musical ideas, the production styles and techniques, and the common sound qualities, present or implied, throughout the album. It is largely influenced by the order of the songs (or musical works) on the recording, the timing of silences between the pieces, and the perceived performance intensity of the pieces.

It is imperative that the album move artistically from one song to another. Little or no space between pieces (suddenly moving to a new work) or lengthy pauses between pieces may be artistically correct, depending on the material, the context of the recording, and the intended overall impression. Some pieces of music end in silence; this time is used for listener reflection, for a sense of drama, or to allow the music to reach its own sense of conclusion. The lengths of silences between pieces must be carefully calculated, to effectively serve the individual pieces of music and the overall project.

Some slight adjustments of loudness between individual works (songs) of an album project are common. These loudness differences are related to, and often aligned with, the perceived performance intensity of the individual pieces. A piece of music that has a perceived performance intensity of forte may have a slightly higher loudness level, on the master, than a piece with a perceived performance intensity of mezzoforte.

The master tape of the entire project must, however, be reasonably consistent in terms of loudness. This will allow for the dynamic relationships of the recordings to be correctly transferred to any other format (CD, television, radio, etc.). In certain applications, dynamic peaks are limited, to allow the project to transfer into other formats (such as broadcast media).

Frequency response of the master may be altered, to allow a recording to sound more accurate on consumer playback and broadcast systems. Often, a recording will be *remastered* for mono formats, instead of equalizing an existing master. This use of remastering also retains the dynamic level relationships that will be altered when two channels are summed into a mono signal.

Global signal processing of an entire recording project rarely occurs. Signal processing the overall project may be used when compiling a sound track, or when all the individual master tapes of a project have the same (or similar) technical deficiencies. Altering the sound of a recording at this stage of the production process is counter productive; it will often create more problems that it solves.

Technically, the project must have certain characteristics for certain applications, such as a specific equalization for broadcast compatibility. Calibration tones and tape formatting specification information must accompany the tape, to assist in this process.

MUSIC LISTENING: ALTERING THE PERFORMANCE

The final listener of the recording may function as a *secondary recordist.* The final form of the recording may be shaped by the listener, through consciously altering the original characteristics of the recording. The listener may alter the sound qualities of a recording, to align with their own personal preferences and distort the sound qualities that were crafted by the recordist.

Further, the sound reproduction systems of the final listener of the recording may be significantly different than the system that was used as a reference during the production of the recording. The listening environment and equipment used for home playback (sound reproduction) are almost never similar to (let alone the same as) those used in monitoring of the process of recording. The differences between studio and home listening environments, as well as between studio and consumer sound reproduction systems, cause great changes to be made in the sound qualities of the recording during playback in home listening environments.

The listener alters the original sound characteristics of the recording by:

- Using considerably louder or softer playback loudness levels than the original recording;
- Adjusting playback equalization;
- Selection of playback equipment;
- Location of the playback system (especially loudspeakers) in their homes;
- Listening to the recording in spaces that distort the recording: small rooms, rooms with parallel surfaces, automobiles; and
- Listening to the recording on headphones.

The music recording will not have the same characteristics in the home listening environment, as in the recording studio. The recordist will hope that the music recording will not be radically distorted, when it is performed in each individual listener's home. At the same time, the recordist must acknowledge reality: such alterations will take place, to varying degrees, much more often than not.

III

The Evaluation of Sound in Audio and Music Recordings

8

The Process of Evaluating Sound

All people in the audio industry are required to analyze sound. The process of carefully evaluating sound, in one context or another, is an integral part of all positions in the audio industry. Sound must be evaluated and analyzed to perceive and evaluate all aspects of audio. Whether the aspects of audio are equipment functions or media productions, are music productions or the technical quality of a sequence of test tones, sound is being analyzed by the listener. Sound is even evaluated by the final consumers of the audio industry: home listeners will evaluate sound to understand the material of the production and to judge the quality of their playback system.

The previous experience, knowledge, cultural conditioning, and expectations of the listener have a direct impact on the level of proficiency at which the listener is able to evaluate sound. With increased experience in evaluating sound comes increased skill and accuracy and a new set of factors influencing conditioning and expectations. The process of evaluating sound can be learned and greatly refined. It should be a primary objective of all people in the audio industry to be more sensitive and reliable in their evaluations of sound.

Nearly all the sound material evaluated by people in audio is not notated. The sounds are not accompanied by written representations (such as a musical score), to assist the listener in understanding, recognizing, or evaluating the material being heard. Furthermore, no language exists to assist people in audio in their communications "about" sound. In nearly all facets of audio, sounds do not coexist in another form that would allow the other senses or the thought processes of conceptualization to assist the hearing experience during the evaluation of sound.

The process of evaluating sound has been devised to provide a means for evaluating sound in its many forms and uses; it will also provide a vehicle whereby meaningful information about sound may be communicated. Critical listening and analytical listening processes are very similar and are performed side-by-side. The critical listening process describes the perceived physical states of the sound material out of applications contexts, and the analytical listening process describes the perceived physical states of the sound material within the context of the individual musical composition.

FUNCTIONS OF SOUND EVALUATION

Recording engineers and producers, obviously, must have well-developed listening skills, as sound evaluation is one of the most important functions to their job activity. This need for highly refined skills obviously holds true for composers and other musicians, especially those involved in the audio recording processes.

Whenever someone listens to a sound, whether the person is listening to a piece of music for the performance of an artistic idea, listening to the sound qualities of a particular piece of audio equipment, or is listening to the effectiveness of the foley sounds in a motion picture sound track, an evaluation process is taking place.

In fact, as many functions for sound evaluation exist as there are job functions within the multitude of positions in the audio industry and within the many media applications that have been found for audio. These many functions for sound evaluation necessitate a method for evaluating sound that can easily be transferred to a variety of contexts, and yet easily yield meaningful and significant information. The method must transfer between musical contexts and abstract, critical listening applications.

The evaluation process is vital to correctly collecting and interpreting information about the sound object or sound event, and the process is required of the technical, as well as the artistic, people in the industry. All people in audio work directly with some aspect of sound. The aspects of sound that they work with might be vastly different, yet they must communicate directly and accurately.

As seen in Chapter 3, the perception of the same material will yield different information to different listeners (or to the same listener on different listenings), depending on: (1) the type of information the listener is trying to determine from the specific listening process, (2) the listener's past experiences and knowledge, (3) the focus of the listener, (4) the level of attentiveness of the listener, and (5) the social-cultural conditioning of the listener.

The evaluation process must provide the foundation for communicating precise information on the content and quality of sound between these two or more people of widely divergent backgrounds. In order for a concept to be communicated accurately, the concept must be described, using common information. The only information common between humans will be the physical states of the sounds themselves, and their knowledge of how their perception distorts the sound.

The evaluation process must use the physical states and activities of sounds as the only universal, reliable information available for describing and defining sound. The evaluation process will describe the activities and states of the physical dimensions of sound and will define the sound characteristics objectively—using subjective human impressions of interpretation and perception only when appropriate to contexts of meaning.

People have been talking about sound for centuries, without a vocabulary. Audio professionals have a sophisticated understanding and clearly defined and complex ways of utilizing sound, without a way to communicate their ideas about sound. This evaluation process provides a means through which meaningful information about sound can be communicated, and also a point of departure for developing a suitable vocabulary for sound.

PERSONAL DEVELOPMENT AND THE SOUND EVALUATION PROCESS

The skills and thought processes required for evaluating sound must be learned. Developing listening skills and evaluating thought processes require regular, focused, and attentive practice. Patience is required to work through the many repetitions that will be needed to master all the skills necessary to accurately evaluate sound. Each individual will develop at a separate pace, as with any other learning.

The process has much in common with traditional forms of music-related ear training. Many of the skills learned by musicians will transfer to the sound evaluation process; an ability to take traditional music dictation will be of assistance in learning the process of evaluating sound, but is not required. The traditional skills emphasize pitch relationships in musical contexts; this comprises a very small part of the sound of audio. The skills of making time judgements and an awareness of activities in pitch, dynamics, and timbre will need to be developed much further than traditional approaches allow.

The process of evaluating sound will be performed more quickly and accurately, with the development of the listener's auditory memory and ability to recognize patterns in the various aspects of sound. The individual must be conscious of the memory of the sound event and must seek to develop the auditory memory, to sustain an impression of the sound long enough to describe or notate certain characteristics about the sound event.

Auditory memory can be developed. As one learns what to listen for, and as one understands more about sound and how it is used, his or her ability to remember material increases proportionally. This is similar to the process of learning to perform pieces of music by listening to recordings of performances and mimicking the performances. With repetition, this seemingly impossible task becomes much easier. Listeners often remember more than their confidence allows them to recognize; the listener must learn to trust their memory and immediately check his or her evaluation, to confirm the information.

The human mind seeks to organize objects into patterns; sound events have states or levels of activity of their component parts that will often tend to fall into an organized pattern. The listener must be sensitive to the possibility of patterns forming in all aspects of the sound event, to allow greater ease in the process of evaluating sound.

The individual will continue to become more accurate and consistent in evaluating sound, the more he or she practices the skills and follows the processes utilized herein. The development of these skills must be viewed as a long-term undertaking. Some of the skills might seem difficult or impossible, during the first attempts. The reader must remember that previous experiences might not have prepared him or her for certain tasks. The skills are obtainable and desirable. The individual will function at a much higher level of proficiency in the audio industry after having obtained these critical and evaluative listening skills.

The skills will be most readily acquired in the order presented in this book. The reader should return to previously "learned" skills often, to rethink and rework the materials, to improve performance. The reader would do well to commit to the idea that mastering the skills of sound evaluation is a lifelong process—one that should be consistently practiced and evaluated. New controls of sound are continually being developed by the audio industry; these new controls create new challenges on the listening abilities of those in the audio industry.

THE SOUND EVALUATION PROCESS AND THE LISTENING EXPERIENCE

The only resources that the listener has to assist in evaluating sound are knowledge and experience. The listener will rely on his or her immediate and past listening experiences to analyze recorded/reproduced sound. Relating what he or she hears to that knowledge and experience, the listener is able to make observations about and to define the sound material.

The evaluation process will follow the sequence of activities:

1. Perceiving the element of sound, or the activity of material, to be analyzed, at a defined focus or perspective;
2. Recognizing the material;
3. Defining material; and
4. Observing the states or treatments present in the material or its activity, or between the element and the musical context.

Evaluating sound begins with the perception of the sound event or the sound object. The term "sound event" is used throughout this discussion; the reader must bear in mind, this process is also applicable to the "sound object." A sound event can be any sound, aspect of sound, or sequence(s) of sounds that can be recognized as forming a single unit. The event may be at any hierarchical level of musical context or of sound quality analysis, from a distant perspective (such as the shape of the overall piece of music) to a close perspective of a focus on some nuance (such as a small change of the spectral content of timbre), but it must have a defined perspective. Each sound analysis will have its own focus on perspective that must be well defined in the listener's mind.

Next, the listener must recognize and, in some way, identify the sound event. This act is necessary to differentiate the event from the material that precedes it, follows it, or is occurring simultaneously with the sound event. The sound event will have parameters within which it exists and through which it is defined. It will have points in time where it begins and ends, and will be perceived within the musical/communications context or in isolation. It will be defined by the unique states and activities of the components of sound.

Whatever the content of the sound event, the listener must perceive it as a single unit that is in need of definition and is capable of being evaluated. This will be accomplished through identifying and recognizing the boundaries within which the sound event exists.

Third, the sound event is defined by the listener, through scanning his or her perception of the sound event, in immediate and short-term memories. The definition process seeks to compile information on the sound event. The sound event is defined by the activities or unique qualities of the materials that caused the sound event to be recognized as being separate and distinct from the materials that preceded it, succeeded it, and/or occurred simultaneously with the sound event. The definition of this activity is often the most difficult task of analyzing a sound event.

Any number of repeated listenings of the sound event will be needed, to extract all the information pertinent to defining the sound event. The skills required to define the sound event will need to be developed. As the listener acquires greater evaluation and listening skills, the number of listenings required to define a sound event are reduced significantly.

The final step is to make sense of the information that was accumulated through defining the sound event. Meaningful observations are made by comparing the information that defined the sound event to sound events that the listener has previously experienced. The listener will use both long- and short-term memory to compare sound events recently experienced, as well as events that are well known to the listener. These other sound events are evaluated for their relationship to the defined sound event. The listener will be looking for same, similar, and dissimilar states of activity and other attributes in the other known sound sources, as those that defined the sound source, to assist in making meaningful observations about the sound event. The process of evaluating the sound event will be completed when the listener has compiled enough detailed information to make all meaningful observations required of the event and its context.

The sound evaluation process is a clear system of routines. It is directly related to the listening experience. The routines follow the order:

1. Identifying desired perspective, with suitable alteration of the listener's sense of focus;
2. Defining the boundaries of the sound event;
3. Gathering detailed information on the material and activity; and
4. Making observations from the compiled information.

The perception of the individual sound event allows for identifying the proper perspective and focus of the listener. The listener consciously decides the level of detail at which the sound event will be evaluated–the perspective. The listener consciously decides on the center of his or her attention (on what aspect of the sound event that information will be sought), in relation to the set perspective–the focus. The sound event and its component parts can then be identified and isolated from all other aspects of the sound and piece of music. The perspective at which the listener has identified the sound event becomes the reference level of the hierarchy, or framework, for the individual X-Y graph of the sound event.

Next, the sound event will be defined by its boundaries of states and activities. It is most often defined by (1) when it exists (its time line), (2) identifying the most significant sound elements that provide the event with its unique characteristics (the states and values of the components of sound that comprise the sound event, both unchanging and transient), (3) the highest and lowest levels (boundaries) within the characteristics (the extremes of levels of activity to be mapped against the time line, or that do not change over time—values of state), and (4) the relationships of how the sound event's characteristics change over time (amount of change and rate of change of values mapped against the time line).

Defining the sound event will include:

1. Determining the time line: beginning and ending points in time of the event, defining suitable time increments, to allow the activity of the components of sound that characterize the event to be clearly presented.
2. Determining which of the components of sound (within the established level of perspective) hold the significant information that characterize the sound event. These are the components that supply the information that defines the unique characteristics of the sound event, and must be thoroughly evaluated, to understand the content of the sound event.
3. Determining the boundaries of the components of sound that characterize the event. These will be maximum and minimum values found in each of the components of sound.
4. Determining the speed at which the fastest change takes place in the components of sound that comprise the event. This will assist in defining the most suitable smallest time increment of the time line.

The third step is compiling detailed information on the material and activity; it will add detail to the above step. Listing the sources (elements to be analyzed) will draw the listener into the evaluation process quickly and directly, and it should be one of the very first steps in compiling detailed information, for evaluating sound.

The components of sound that characterize the sound event will be evaluated, to determine (1) precise states of value (with the smallest increment between values) of the various elements (such as pitch, dynamic levels, etc.), and (2) placement of those states of value against the time line (mapping the activity of the component parts of the sound event). The components of the sound will be closely evaluated, with as much detail as possible, to determine their precise states of value.

Most often, the components of the sound event are transitory (change over time), and must have their values related to a time line. This information will be plotted on a two-dimensional graph (in the next section), thus allowing the information to be written, to assist in this evaluation process and to make the information available for future use. This process involves all the skills of taking music dictation. In fact, it is a type of music dictation for some new, and some previously ignored, aspects of sound.

A written representation (of the states of value and activities of the components of sound in the sound event) will be created through the process of following steps 1 through 3. Making a written representation of the sound event will make it much easier to compile information on the sound event. This process will allow the listener to check previous observations for accuracy, to focus on particular portions of the sound event, and to continue gathering information on the sound after the sound has stopped.

The final activity in evaluating the sound event is one of making observations from the compiled information. The type of observations made will vary considerably, depending on context and the sound itself.

For example, if the observations being made concern the functioning of a particular piece of audio equipment, the evaluations will center around the aspects of sound that the particular piece of equipment acts upon. Observations will be focused on the effectiveness of the piece of equipment, the integrity of the audio signal, and any differences between the input and the output signals.

The listener/evaluator will formulate questions and will use the data compiled in the above steps, to answer those questions. What questions to ask are determined by their appropriateness to the purpose of the evaluation. The answers produced through this process will be ones of substance and will be directly related to the sound event; they will not produce subjective impressions or opinions.

The observations made in this final evaluation process need not be profound in order to be significant. Often, the simplest, most obvious observations offer the most significant information concerning a sound event or sound object.

GRAPHING VALUES AND ACTIVITY

Creating a written representation of the sound event will greatly assist the listener in understanding the sound, communicating about the sound, evaluating the sound, remembering the sound, and recreating the sound. While the reader will not seek to perform a written evaluation of every sound event presented to him or her, the process of performing a detailed evaluation of the sound event will provide information that would otherwise go unobserved. Creating written representations of the sound material is required for finely developing sound evaluation skills; it will also provide the reader with a useful resource, to assist in evaluating sound.

The traditional two-dimensional line graph is quickly understood, easily designed, and readily completed by most people. Therefore, it has been selected as the basis for notating (creating written representations of) sound events.

The line graph will nearly always be used with time as the horizontal (X) axis. In this way, values of states of the component parts of the sound event can be plotted, with respect to time. This allows the sound event to be observed from beginning to end at a glance, out of real time.

The length of the sound event or sound object that can be plotted on a single graph, is dependent upon the selected increments of the time axis, or *time line* (Fig. 8-1). Events of great length (and little detail) may be plotted on a single graph, and events of short duration (and great detail) may be plotted on a single graph.

A balance must be found when selecting the appropriate time increment for the time line. The sound event must be easily observed in its totality (from beginning to end), and the graph must have sufficient detail of information, to be of use in observing the qualities of the event.

Time increments will be selected for the X axis that are appropriate for the sound event. Time increments will take one of two forms: (1) units based on the second (millisecond, tenths of seconds, groups of seconds, etc.), and (2) units based on the metric grid (individual or subdivisions of pulses, measures, groups of measures). If the sound material is in a musical context, the metric grid will nearly always be the preferred unit for the time axis. Humans judge time increments more accurately with the recurring pulse of the metric grid acting as a reference.

In general, when the sound evaluation utilizes the metric grid, a process of analytical listening is occurring. The critical listening process most often uses real time increments, and not the metric grid. The difference is one of context and focus.

If the sound material being evaluated is not in a musical context, increments based on the second must be used. It will be common to use

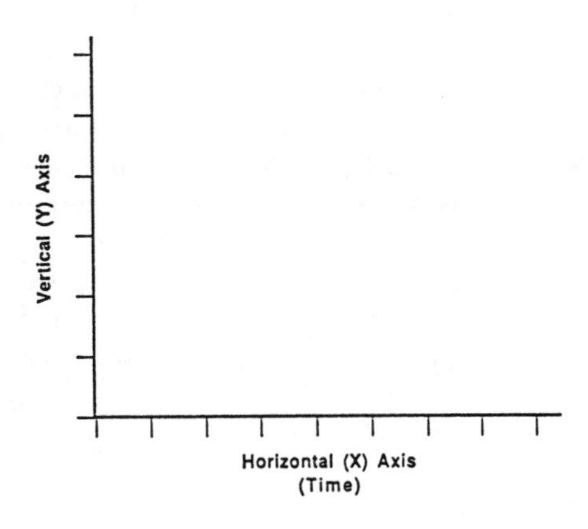

FIGURE 8-1. X-Y Line Graph.

increments based on the second when evaluating timbre relationships (including sound quality and environmental characteristics). While conceptualizing the pulse of M.M.=60 (or an integer or a multiple thereof) will provide some reference to the listener in making time judgements without a metric grid, this activity may not always be appropriate. It may distort the listener's perception of the material, and the reference may be unstable, as the listener's attention will rightly be focused elsewhere.

A stopwatch will often assist in evaluating larger time units (to the tenths of seconds). The ability to judge time relationships can be developed. Practicing the following exercise will, over time, allow the reader to refine their skills in accurately making time judgements.

With practice, the listener will develop the ability to make accurate time judgements of a few milliseconds, within the context of known, recognizable sound sources and materials (Exercise 8-1). This skill will be invaluable in many of the advanced sound evaluation tasks regularly performed by audio professionals.

The time unit used in the line graph will be that which is most appropriate for the sound event. The time increment selected must allow the graph to depict the sound event accurately; the smallest perceivable change in the components of sound being analyzed must be readily apparent; yet, as much material as possible should be contained on a single graph.

The components of sound to be plotted and the boundaries of states and activities of those components are next determined. In the two initial stages of the sound evaluation process, the listener determines those components of the sound event that provide it with its unique character. These components of the sound event will be those selected for analysis, through plotting on the line graph.

The component of the sound event to be evaluated will be placed on the vertical (Y) axis of the line graph. Through the second step of the sound evaluation process (discussed in the previous section), the listener will determine the maximum and minimum values found in each of the components of sound. These maximum and minimum values will be slightly exceeded, when establishing the upper and lower boundaries of the Y axis (depicting the component).

Exceeding the perceived limits of the values of the components allows for errors that may have been made during initial judgements of the boundaries, and allows for greater visual clarity of the graph. Boundaries should be exceeded by 5 to 15 percent, depending on the context of the sound event and the space available on the line graph.

Next, the minimum changes of activity and levels are determined. Through "Step 3" of the sound evaluation process (outlined in the previous section), the listener will determine the smallest increment of

Time Judgement Development Exercise

Using a digital delay unit and a recording of a high-pitched drum sound (or a drum machine):

1. Route a non-delayed signal to one loudspeaker, and a delayed signal to the other loudspeaker.
2. Delay the signal, perform many repetitions of the sound, using time-increments that are easily recognizable (always move by the same time-increment unit, such as 100 ms, within a given listening session).
3. As confidence is obtained in being able to accurately judge certain time units increases, move to smaller and larger time units, and repeat the procedures of step 2.
4. When control of time relationships is accurate within certain defined limits and the reader is confident in this ability, test that accuracy by routing both the direct and delayed signals to both loudspeakers (or to a single loudspeaker).
5. Continue to work through many repetitions of time increments in a systematic manner, comparing the qualities of the time relationships of each listening, to previous and successive material, in a logical sequence (a suggested pattern or sequence: 150 ms, 125 ms, 100 ms, 75 ms, 125 ms, 150 ms).
6. Continue moving to smaller and smaller time units, until consistency has been achieved at being able to accurately judge time increments of 3 to 5 ms.

EXERCISE 8-1. Time Judgement Development Exercise.

value for the components of the sound event. This smallest increment of value will serve as the reference in determining the correct division of the Y axis. It is necessary for the Y axis to be divided, to allow the smallest value of the component of sound to be clearly represented on the vertical axis of the graph, as was the concern with dividing the X (time) axis of the graph.

The division of the vertical axis must allow the graph to depict the sound event accurately. The smallest significant change in the components of sound being evaluated must be immediately understood by the reader of the graph, and yet the vertical axis must not occupy so much space as to distort the material. The reader of the graph must be able to identify the overall shape of the activity, as well as the small details of the activity of the component that the graph represents. A balance between limitations of space and clarity of presentation of the materials must always be sought.

It is not always desirable for each component of the sound event to have its own line graph. Many times, several components of a sound event can be included on the same graph and plotted against the same time line. *Multi-tier graphs* allow several components to be represented against the same time line, with the advantage that their characteristics will be more easily interrelated during the final formulation of observations.

The vertical (Y) axis of the line graph is divided into segments. Each segment is dedicated to a different component of the sound event and will have its own boundaries and increments. Plotting a number of components of sound against the same time line not only makes efficient use of space on the graph, it allows a number of the characteristics of the sound event (perhaps the entire sound event) to be viewed simultaneously. By placing a number of the components of the sound event against the same time line, it is possible to give a more complete and more easily understood representation of the sound event.

The person reading the graph will be able to extract information more quickly from a multi-tier graph that from a series of individual graphs. In addition, plotting various components of the sound event against the same time line lets a person compare the states and activities of the various components of the sound event, in ways that would be more difficult (if not nearly impossible) were these components separated.

Multi-tier graphs of a specific nature will be used for certain advanced evaluation processes. In those cases, the graphs will always appear in a predetermined format.

Multiple sound sources within the same component of the sound event must also be graphed (Fig. 8-2). It is quite common for more than one aspect within a component of the sound event to be taking place at any one time (such as the sound of harmonics and overtones within the spectrum of a sound). This activity would require a separate tier of a multi-tier graph for each sound source in each component of the sound event being evaluated. The line graph would quickly become large and unclear.

As long as the segment of the graph can remain clear, it is possible for any number of sound sources to appear on any graph. When more than one sound source appears against the same two axes, the sound sources must be differentiated.

Sound sources may be differentiated in a number of ways. Each of these ways may be useful, depending on the resources of the reader, the nature

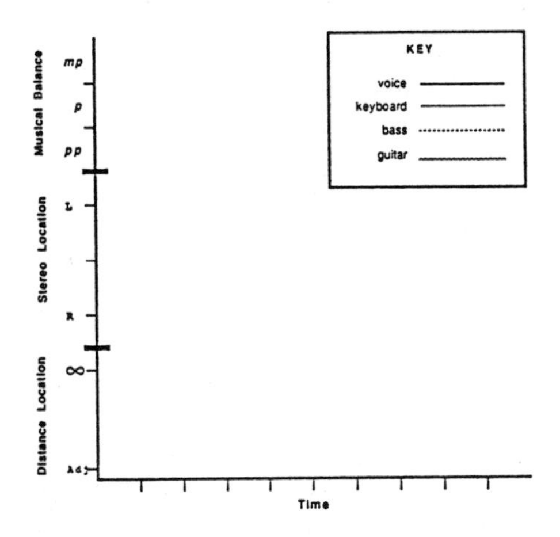

FIGURE 8-2. Multitiered Graph.

of the sound event, and the context of the sound event. The lines that denote each sound source may be labelled. Labelling lines is accomplished by placing a "number" or the "name" of each sound source near the appropriate line on the graph. This type of differentiation is useful for graphs that contain only a few sound sources.

Providing a different line configuration for each sound source is often a suitable way of differentiating a number of sources on the same tier of an X-Y graph. Combining dots and dashes, or inserting geometric shapes into the source lines, are useful for differentiating up to about eight sound sources on the same graph.

Sound sources are most easily differentiated by adding a variety of colors to the graph. Assigning a different color to each sound source is the clearest way to place a large number of sound sources on the same graph. The number of sources that could be placed on the same graph is limited only by the number of easily recognized colors available.

Using different colors has the further advantage of being able to define groups of sound sources by assigning a color to the group and assigning a different line configuration (combination of dots and dashes) to the individual sound source.

Using color is not always feasible, but it is the preferred method of placing a number of sound sources on the same graph. Utilizing varied and distinct line configurations for each sound source is the next most flexible and clear method of differentiating sound sources. Combining color and line configurations will produce the most organized and most useful graphs. Individual sound sources must always be easily distinguished on line graphs; readily identifiable lines that have been precisely defined (in a key) will ensure the clarity and usefulness of the graph.

The same sound sources may be depicted on a number of tiers on a multi-tier graph. In this case, care must be taken to define each sound source and to depict the sound sources in the same way on each tier (either by the same name, color, or line configuration). This will allow the reader of the graph to quickly and accurately determine the states and activities of all the sound sources (or aspects of the sound sources) over time. A key of the sound sources plotted should be created to ensure this clarity.

Varying line thicknesses as a means of differentiating sound sources by providing each source with a line of a different thickness should *not* be used. This technique will obscure the information of the graph. Varying line thicknesses will cause the sound to visually appear to occupy an area of the vertical axis (a state that is only accurate for a few select components of sound).

A ***key*** lists sound sources of the sound event, coupled with a chart of how the individual sound sources are represented on the line graph (see Figure 8-2). This listing of sound sources, with their designations, must be included in each line graph that contains more than one sound source (unless the lines are labelled).

The act of listing sound sources takes place as one of the first activities to be undertaken in the entire process of evaluating sound. Sound sources are the individual elements of activity within the level of perspective that is the focus of the sound evaluation. A listing of the sources (elements to be analyzed) will draw the listener into the evaluation process quickly and directly, and it should be one of the very first steps of the evaluation.

Plotting the individual sources against the time line, without concern as to states of value and activity of the various component parts of the sound event, will allow the listener to compile preliminary information on the material, without getting overwhelmed by detail. This process is an excellent first step in getting acquainted with the activity of writing down material that is being heard (taking dictation). It may become a common initial activity, each time the listener undertakes a detailed evaluation of a sound event. The ability to perform the following exercise will be assumed throughout the remainder of the book, as this process will be repeated, at least conceptually, before almost all future exercises.

As a common activity, the reader should (1) create a listing of the sound sources of a sound event, (2) create a time line of the event, and (3) plot the defined sound sources against the time line. The reader should practice Exercise 8-2, and become comfortable with the process.

Figure 8-3 provides an example of placing sound sources against a time

Exercise in Mapping Sound Sources Against Time Line

Graph the first few major sections of a piece of popular music through the following steps:

1. Compile a listing of all the sound sources of a piece of music (individual percussion sounds and vocal parts must all be listed separately).

2. Create a suitable time line, by:
   a. determining the pulse (metric grid) of the piece of music;
   b. mapping the pulses into groups of strong and weak pulses (measures); and
   c. plotting those divisions of time (measures) on the horizontal (X) axis in suitable increments to depict the example being graphed.

3. Plot the individual sound sources against the time line. Each sound source will have its own location along the vertical axis, making it unnecessary to make distinctions between the lines of each sound source on the graph.

EXERCISE 8-2. Exercise in Mapping Sound Sources Against Time Line.

line. The listener will be able to reference the figure against The Beatles' recording: "She Said She Said." The recordist must be able to quickly recognize sound sources and focus on their activity. As an exercise, complete the graph by plotting the presence of the high hat part against the time line.

COMPLETING THE PROCESS

Graphs will often need to be supplemented by verbal descriptions and evaluations of the sound event. Simply graphing the sound event does not complete the evaluation process. The contents of the graphs need to be described, through observations of the content and activity of the sound sources. In all instances, the language and concepts used to define and describe the sound event must be completely objective in nature. All descriptions refer to the actual states of existence and the changes in states,

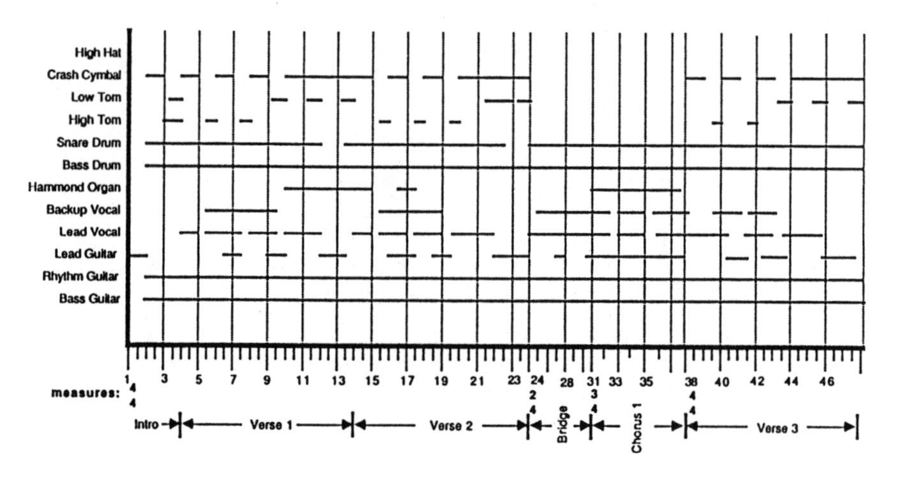

FIGURE 8-3. Placement of Sound Sources Against Time Line—The Beatles: "She Said She Said."

of the components of the sound event. The descriptions lead to overall observations about the qualities of the sound event itself and its relationships to the context in which it is presented, or the sound object itself.

The listener must never use subjective impressions or descriptions of the sound event, when completing the evaluation process. Such impressions are unique to each individual, cannot be accurately communicated between individuals (they mean something different to all people), and do not contribute to understanding and recognizing the characteristics of the sound event. Subjective impressions or descriptions do not contribute pertinent, meaningful information about the sound. Evaluations or descriptions that are based on an individual's personal perceptions or impressions about a sound are meaningless for communicating concepts and ideas, and will not contribute to the understanding of the characteristics of sound. They have no place in the evaluation process.

9

The Evaluation of Pitch in Audio and Music Recordings

Many analytical systems have been devised to explain pitch relationships in music. These theories about music are sets of evaluation criteria that vary considerably in content. Any particular analytical system may or may not be suitable to evaluate the music under consideration. The recordist must be able to recognize which analytical system is appropriate for studying the pitch relationships of a particular piece of music, if these evaluations are expected of them.

Consideration of the many available systems of pitch analysis is well outside the scope of this book. These theories about music all have in common their attempt to explain the analytical listening experience and to extract basic information about the structure and form of the music.

Information about the artistic element of pitch levels and relationships will be related to (1) the relative dominance of certain pitch levels, (2) the relative registral placement of pitch levels and patterns, or (3) pitch relationships: patterns of successive intervals, relationships of those patterns, and patterns and relationships of simultaneous intervals.

Nearly all musical instruments were specifically designed to produce many precise variations in pitch, far fewer variations in timbre, and a continuously variable range of loudnesses. Most Western music places great emphasis on pitch information, for communicating the musical message. Pitch is the central artistic element of most music; the other artistic elements will most often support the activities occurring in pitch.

Pitch changes are perceived in both analytical and critical listening contexts. The critical listening process will transfer the perceived pitch information into frequency estimation. The analytical listening process will relate the pitch relationships to the musical context and the musical message. The reader will be developing ways of analyzing musical sounds that will also develop critical listening skills. The same aspects of sound are being heard in each process, with the context and focus being different.

The pitch analysis methods presented in this chapter are concepts that are of particular importance to sound recordings. The recording process has given the creative artist new controls over pitch and pitch relationships; these methods provide a means to define those concepts. The information gathered by these analysis methods will assist in understanding the piece of music (analytical listening) and in evaluating sound quality (critical listening), depending on how the information is applied.

ESTIMATION OF PITCH LEVEL

The processes of critical listening and analytical listening define perceived pitch differently. Frequency estimation through pitch perception will be used throughout the evaluation process, to allow for critical listening observations. *Pitch/frequency registers* will be used to estimate the relative level of the pitch material and to allow the information to be directly transferred between the two contexts.

Frequency estimation is a fundamental skill that must be developed by the audio professional. The use of pitch/frequency registers will assist the reader in defining perceived pitch and frequency levels. The registers will serve as reference areas and will provide a basis for describing perceived frequency and pitch levels. The registers for estimating pitch level in relation to pitch and frequency will be used in many of the evaluation processes that follow, and should be committed to memory, to provide meaningful reference levels.

The ranges of human singing voices stretch from the *low–mid* and *mid* registers (male voices) to the mid and *mid–upper* registers (female voices). Most musical activity occurs in the mid and mid-upper registers; this is where many instruments sound their fundamental frequencies, and where most melodic lines and most closely spaced chords are placed in musical practice. (Figs. 9-1, 9-2, and 9-3.)

The sibilant sounds of the human voice occur primarily in the *high* register. Within the *very high* register, humans have the ability to hear nearly two and one-half octaves. While this register is not playable by acoustic instruments and by human voices, much spectral information is often in this register.

The reader should work through the following exercise to begin developing skill at estimating pitch levels. This skill is central to many of the evaluations commonly performed by all people in audio, and will be used throughout the remainder of this book. The listener can easily transfer perceived pitch level into frequency estimation, by using the above registers. The reader should practice transferring various pitch levels to frequency, and the reverse.

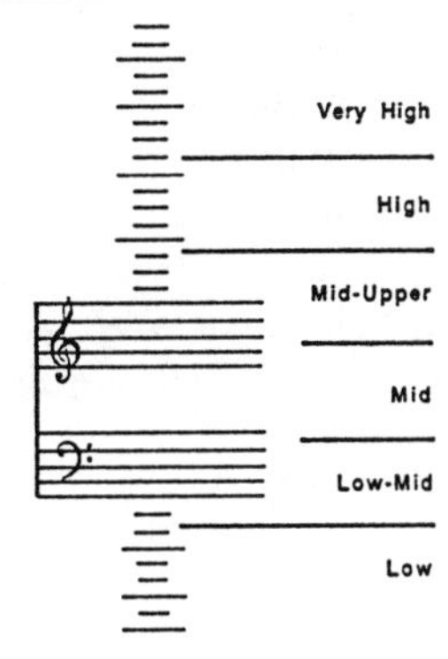

FIGURE 9-1. Registral Designations.

Register	Pitch Range	Frequency Range
Low	up to C2	up to 65.41 Hz
Low-Mid	D2 to G3	73.42 to 196 Hz
Mid	A3 to A4	220 to 440 Hz
Mid-Upper	B4 to E6	493.88 to 1,318.51 Hz
High	F6 to C8	1,396.91 to 4,186.01 Hz
Very High	C8 and above	4,186.01 and above

FIGURE 9-2. Registers for the Estimation of Pitch Level in Relation to Pitch and Frequency.

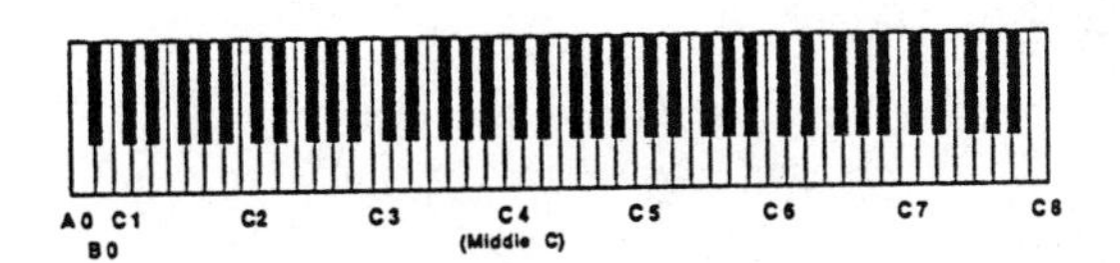

FIGURE 9-3. Octave Divisions of Keyboard.

It is possible for the experienced listener to consistently estimate pitch/frequency level to within an interval of a minor third (one-quarter of an octave), after considerable practice and experience (Exercise 9-1). Within several weeks of concerted effort, the reader should be consistent within a perfect fifth (a bit over one-half an octave), with accuracy continuing to increase with regularly applied effort. Developing listening skills is a long-term project.

PITCH AREA AND FREQUENCY BAND RECOGNITION

Many sounds occupy a *pitch area*, not a specific pitch. These sounds are perceived in musical contexts, according to an area of pitch level between two boundaries; the boundaries may, at times, be unstable in dynamic level or in pitch level, and difficult to precisely define at specific levels. The sounds are defined in terms of (1) the density (amount) of the pitch information within the pitch area, (2) the width of the pitch area (the distance between the two boundary pitches), and (3) primary and secondary pitch areas.

Some sounds will have several separated pitch areas. The different pitch areas of the sound will be at different dynamic levels and have different densities (the amount and closeness of spacing of the pitch information

Pitch Level Estimation Exercise

Working with a keyboard instrument or a tone generator, practice listening to the boundaries of the registers, for pitch/frequency estimation.

Once confidence has been established in recognizing the general areas encompassed by the registers, begin placing individual pitches against the pitch registers. The listener must rely solely on his or her memory of the pitch registers, in making pitch level judgements.

This listening exercise is most useful when another person performs the pitches for the person estimating pitch levels.

EXERCISE 9-1. Pitch Level Estimation Exercise.

within the pitch areas). One pitch area will dominate the sound, and be the primary pitch area; the other pitch areas will be secondary pitch areas.

Percussion sounds are the most common examples of sounds that occupy a pitch area. The pitch area of these sounds must be defined, to understand their relationship to (1) the other pitch material in the music and (2) the density of the overall musical texture (textural density). The pitch areas of percussion sounds (especially drums) are often closely associated with the other pitch material (and their sound qualities) of the composition.

The process of defining pitch areas should move through the sequence:

1. Determine the pitch area of most activity by first defining the lowest and then the highest boundary of the area (a steep filter may be helpful in determining these boundaries during beginning studies).
2. Determine the area of next highest activity (defined by width, density, or dynamic prominence of the pitch area), by identifying the lowest and then the highest boundary.
3. Repeat the process for all other pitch areas present (specific frequencies/pitches in the spectrum may be audible, despite the sound not having an audible fundamental frequency; these should be identified and plotted).
4. Compare densities between the pitch areas.

The information is plotted on a *pitch area analysis graph*. The graph incorporates the following:

1. The registral designations described above for the "Y" axis, distributed to complement the characteristics of the musical example.
2. A space on the "X" axis of the graph is dedicated to each sound plotted (since this process does not use a time line).
3. The pitch areas are boxed off in relation to the two axes.
4. The densities of the various pitch areas of each sound, relative to one another, are designated by different densities of shading within the area boxes, and a short verbal description using a relative scale from "very dense" to "very sparse."
5. A verbal description (providing objective information related to dynamic levels) should accompany the graph and articulate any important observations concerning the relative dynamic levels that will cause some pitch areas to be more predominant than others (dynamics will be covered in the next chapter).

Figure 9-4 presents the pitch areas of the prominent tom drum sound found in the introduction section of Phil Collins' "In the Air Tonight." The

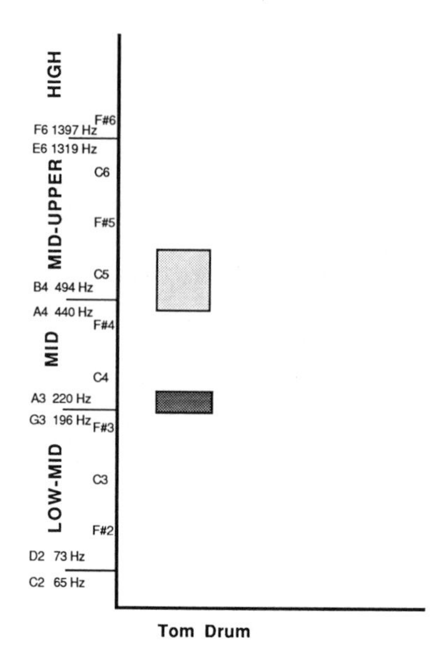

FIGURE 9-4. Pitch Area Analysis—Tom Drum Sound—Phil Collins: "In the Air Tonight."

percussion sound is comprised of two pitch areas. The density of the lower pitch area is moderately dense, and the higher pitch area is rather sparse; the lower pitch area is moderately louder than the higher pitch area. It is interesting to note that the two pitch areas are harmonically related: the lower boundaries of each area are separated by an octave.

The pitch area graph and the objective descriptions of density and dynamic relationships of the pitch areas provide much useful and universally perceived information about the sound. Meaningful communication about this sound is possible with this information.

Additional percussion sounds are plotted in Figure 9-5. As an exercise, the reader should listen to the recording and attempt to determine the dynamic relationships and densities of the frequency areas that have been identified in The Police's work "Every Breath You Take." The various shadings of the boxes provide implied densities.

The reader will be able to apply the skills and concepts of pitch area to recognizing *frequency bands* in critical listening applications. The information for evaluating the states and activities of frequency will be deduced from the perceived pitch information. The skills gained through recognizing pitch areas and frequency bands will later be directly applied to evaluating timbre and sound quality. In fact, the earlier pitch area analyses are rudimentary timbre analyses.

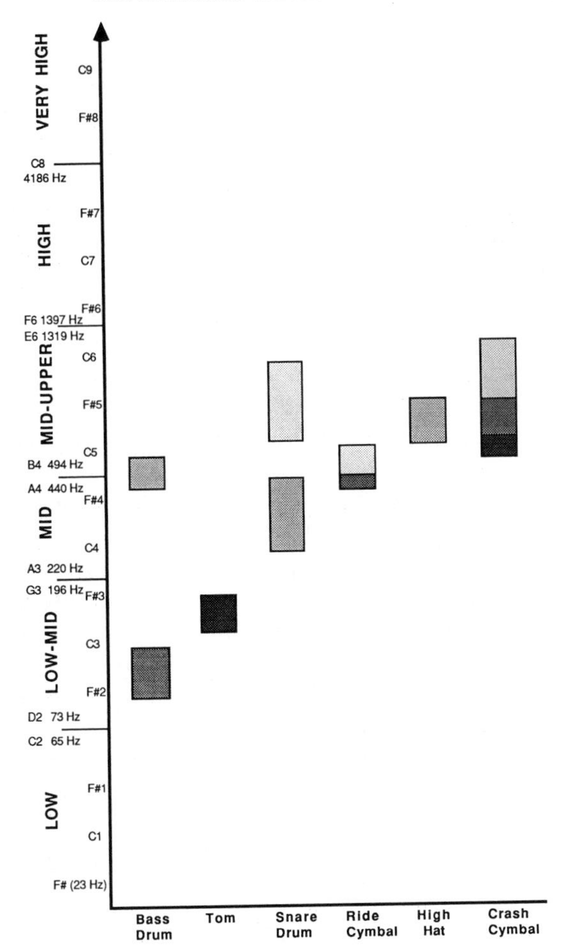

FIGURE 9-5. Pitch Area Analyses—Percussion Sounds—The Police: "Every Breath You Take."

TEXTURAL DENSITY

The process of evaluating textural density is directly related to pitch area analysis. Textural density is the amount and placement of pitch-related information, within the overall pitch area of the musical texture. It is comprised of the pitch areas of the musical materials and sound qualities of the individual sound sources (voices, instruments, or groups of instruments and voices).

The concept of textural density allows each sound source to be perceived as having its own pitch area in the musical texture. It allows for the pitch area that spans human hearing (and the musical texture) to be conceived as a space. Within this space, sound sources appear to be layered according to the frequency/pitch area they occupy. In this form, the concept of textural density is often applied to the processes of mixing musical ideas and sounds in recording production. The pitch areas in textural density are determined in a way similar to defining the pitch areas of non-pitched sounds. The material being plotted is, however, more complex and the process more involved.

First, the musical texture must be scanned, to determine the musical ideas present. The musical ideas might be a primary melodic (vocal) line, a secondary vocal, a bass accompaniment line, a block-chord keyboard accompaniment, or any number of different rhythmic patterns in the percussion parts. The number and nature of possible musical parts are limitless.

Second, the musical ideas must be clearly identified with the sound source or sources performing it. This may or may not be a simple process. It is possible for a single instrument to be presenting more than one musical idea, such as a keyboard presenting arpeggiated figures in one hand and block-chords in another. It is possible for many instruments to be grouped to a single musical idea, and then suddenly have one instrument emerge from the ensemble to present its own material. The possibilities are much more complex than merely labelling each musical part with the name of an instrument, although this is often the case.

Finally, each idea will have its pitch areas defined by a composite of the pitch area of the fused musical material and the primary pitch area of the instrument(s) or voice(s) that produced the idea.

Musical materials are a single concept or pattern. These materials will fuse in a listener's memory and perception as a group of pitches, which occupy a pitch area. The pitch area of the musical material is defined by its highest and lowest pitches, as the boundaries. Within the boundaries, density is created by the number and spacing of the different pitch levels that comprise the musical material.

The pitch area of the sound source will also influence the pitch area of the musical idea. The primary pitch area of the sound source will often be the fundamental frequency of an instrument or voice, perhaps with the addition of a few prominent lower partials. A primary pitch area may also contain environmental cues information (delayed and reverberated sound) that may add density to the sound, without adding more pitch information, pitch-shifted information that may provide additional spectral information and added density, and/or any number of other alterations to the sound source that might provide more pitch information to be present in the primary pitch area of the sound source.

Often, the primary pitch area of the sound source is the fundamental frequency alone. This is especially true when an instrument or voice is being performed at a moderate to low dynamic level. As the dynamic level of a sound source increases, lower partials will often become more prominent, and the width of the pitch area will tend to widen.

The pitch area of each musical idea must first be determined by defining the length of the idea. It is then possible to define the lowest boundary of the pitch area and then the highest boundary of the area. This pitch information is determined by calculating the melodic and harmonic activity of the musical idea (to determine the highest and lowest pitch levels) and then adding detail that pertains to the sound qualities of the sources performing the idea.

The lowest boundaries of pitch areas are most often the fundamental frequencies of the sound sources performing the musical material (roughly following the melodic contour or the lowest notes of a chord progression). The upper boundary is either the highest pitch level of the musical material, or is the top of the predominant pitch area of the highest sound source (and pitch level) of the musical idea. The boundaries of pitch areas of musical ideas may or may not change over time.

The pitch area of a musical idea is the composite of all the appearances of the sound source performing all the pitch material within the time period of the musical idea (the sum of all of the pitch levels, and the pitch areas of the primary pitch areas of the sound sources).

The individual sound source is conceived as a single pitch area, and is identified by boundaries that conform to the composite pitch material. The relative density of the idea is determined by the amount of pitch information generated by the musical idea and the spectral information of the sound source(s). The process of determining pitch area is repeated for each individual musical idea.

Pitch areas may be mapped against a time line. If a time line is used, changes in pitch area as the musical idea unfolds in time can be presented (as appropriate to the musical context). Mapping items against a time line is covered in the following section.

All sound sources are plotted on the *textural density graph*. The graph may take two forms: (1) it may simply plot each sound source's pitch area against one another (as in the pitch area recognition graph), or (2) the sound sources may be plotted against a time line of the work, allowing the graph to visually represent changes in the textural density (the registral distribution of sounds), as the work unfolds. In either graph, (1) each sound source is represented by an individual box, denoting the pitch area (boxes will be of different colors to differentiate sources if they overlap in the second graph above), (2) the "Y" axis is divided into the appropriate registral designations, and (3) the density of the pitch areas should be denoted on the graph, through shadings of the boxes, and verbal descriptions should accompany the graph, as necessary.

The pitch areas of all the sound sources may be compared to one another, as well as to the overall pitch range of the musical texture. Textural density allows the pitch area of all sound sources to be compared. Thus, recordists are able to better understand and control the contribution of the individual sound source's pitch material to the overall musical texture, and of the mix.

The textural density of the beginning of Kate Bush's "This Woman's Work" appears in Figure 9-6. The work utilizes textural density and the registral placement of sound sources and musical ideas, to add definition to the musical materials and sections of the music. Textural density itself helps create directed motion in the music. The musical ideas are precisely placed in the texture, allowing for clarity of the musical ideas. The expansion and contraction of registral placement and textural density of the musical ideas (and sound sources) add drama, to support the musical materials.

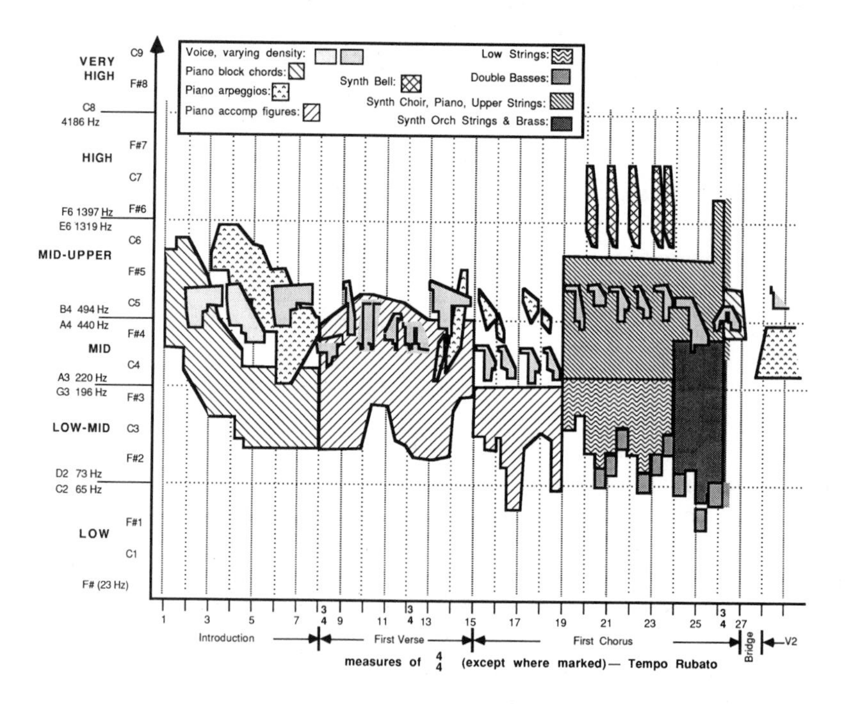

FIGURE 9-6. Textural Density Graph—Kate Bush: "This Woman's Work."

MELODIC CONTOUR

Plotting melodic contour will assist the recordist in a number of ways. First, at the early stages of the recordist's development, graphing melodic contour will help develop skills in pitch/frequency estimation against time. These same skills will later be used in a much more detailed (and initially more difficult) context for evaluating pitch-related information in timbre and environmental characteristics analyses.

Practicing the plotting of melodic contours against the time line will be productive in developing the skills of recognizing pitch levels, of time unit and rhythm perception, and of mapping pitch contours. These skills directly transfer into many of the listening functions of the audio professional.

Second, recognizing the contour, or shape, of the melodic line is important to understanding certain pieces of music or musical ideas. In certain pieces of music, the contour of a melodic line is perceived, instead of the individual intervals. When the melodic line is performed very rapidly (as is easily accomplished with technology), the perception of the line fuses into an outline or shape. The series of intervals that comprise the melodic line are not perceived. The contour of the melodic line is instead perceived.

The melodic contour graph allows the contour of the musical idea to be recognized and evaluated. The shape of the melodic contour may be compared to the shapes of the activities of the other artistic elements, or the melodic material can be appreciated by the qualities of its contour.

The process for defining the activities of all the artistic elements as they relate to a time line (in a musical or critical listening context) will follow the sequence presented below. This process will be modified only slightly for each particular artistic element (or perceived parameter), and the reader should become familiar with the order of activity:

1. During the first hearing(s), listen to the music under evaluation, to establish the length of the time line. At the same time, notice prominent instrumentation and activity of the artistic element (or perceived parameter) that is being evaluated against the time line.
2. Check the time line for accuracy, and make any alterations. Establish a complete list of sound sources (instruments and voices), and sketch those sound sources against the completed time line.
3. Notice the activity of the artistic element for boundaries of levels of activity and speed of activity. The boundary of speed will establish the smallest time unit required in the graph, to accurately present the smallest significant change of the element; the boundary of levels of activity will establish the smallest increment of the "Y" axis required to

plot the smallest change of the artistic element. This step will establish the perspective of the graph, which is the graph's level of detail.
4. Begin plotting the activity of the artistic element on the graph. First, establish prominent points of reference within the activity; these reference points might be the highest or lowest levels, the beginning and ending levels, points immediately after silences, or other points that stand out from the remainder of the activity. Use the points of reference to judge the activity of the preceding and following material; alternate focus on the contour, speed, and amounts of level changes, to finish plotting the activity of the artistic element. The evaluation is complete when the smallest significant detail has been perceived, understood, and added to the graph.

Many listenings will be involved for each of the above steps. Each listening should seek specific new information and should confirm what has already been noticed about the material. Before listening to the material, the listener must be prepared to extract certain information (attention focused at a specific level of perspective), to confirm their previous observations and to be receptive to new discoveries about the example. The listener should check his or her previous observations often, while seeking new information.

Listening to only small, specific portions of the example may assist certain observations at certain points in the evaluation process. In these situations, the listener should intersperse listenings of the entire example, with rehearings of small sections, to be certain that consistency is being maintained throughout the evaluation.

The reader should very rarely write observations while listening. Instead, the reader should concentrate on the material and attempt to memorize his or her observations. This activity will develop auditory memory and will ultimately reduce the number of hearings required to evaluate sounds.

One must first recognize the characteristics of the sound, before they can be written down. This process is about recognizing what is heard, and making a written record of the experience.

The information is plotted on a *melodic contour graph*. The graph incorporates the following:

1. The pitch area registral designations for the "Y" axis, distributed to complement the characteristics of the musical example.
2. The "X" axis of the graph is dedicated to a time line that is devised to follow an appropriate increment of the metric grid or of real time (depending on the material and context).
3. Each sound source is plotted as a single line, against the two axes (the melodic contour is the shape of this line).

4. If more than one sound source is plotted on the same graph, a key should be devised and presented.

Consider the melodic contour of the trumpet material that opens Phil Collins' "Hang In Long Enough" (Fig. 9-7). The melodic line is performed too quickly to be heard as a melody comprised of intervals; instead, this type of fast melodic gesture fuses into a shape, or contour. As an exercise, check Figure 9-7 for accuracy of shape, detail, and registral placement, and determine the tempo of the passage.

Melodic gestures of this type are very common in music. They commonly appear in works, from heavy metal guitar solos to eighteenth- and nineteenth-century keyboard music (especially in the works of Chopin and Liszt, and works in the Rococo style). Transcribing fast melodic passages into melodic contour graphs will provide the reader with practice in developing the following skills required of recordists: pitch recognition (estimation), recognizing and calculating pitch changes, and placing pitch changes against a time line.

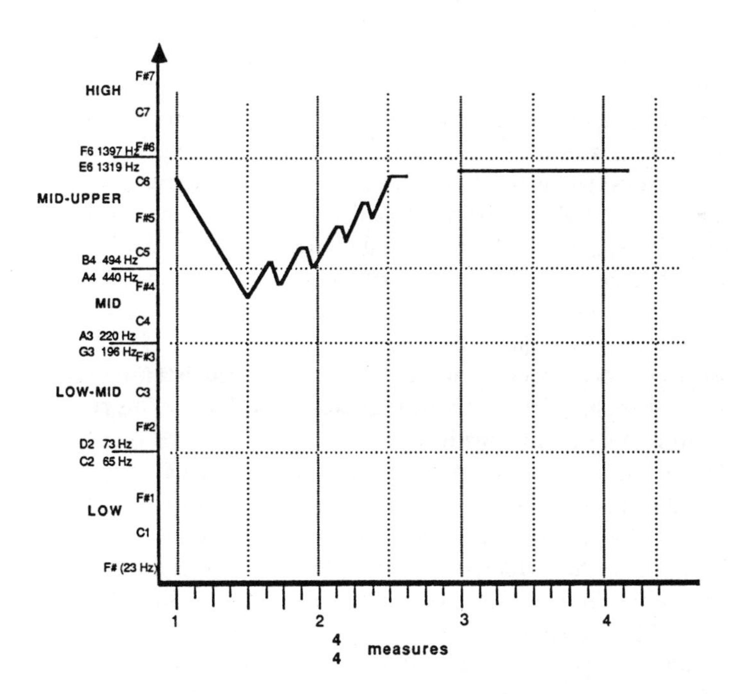

FIGURE 9-7. Melodic Contour Graph–Phil Collins: "Hang in Long Enough."

10

Evaluating Loudness in Audio and Music Recordings

Loudness has traditionally been used in musical contexts, to assist in the expressive qualities of musical ideas. This function of dynamics helps shape the direction of a musical idea, delineate the separate musical ideas (usually in relation to their importance to the musical message), assist in creating nuance in the expressive qualities of the performance, and/or add drama to the musical moment. In all these cases, dynamics have functioned in supportive roles in communicating the musical message.

The recording process has given the recordist more precise control over dynamics than was possible in live, acoustic performance. This increased control has given audio recordings new relationships of dynamics, as well as the potential of placing more musical importance on dynamic relationships. An example of a new relationship of dynamics is the occurrence of contradictory cues between the loudness level at which a sound was performed during the initial recording (tracking) and the dynamic level at which the sound is heard in the final musical texture (the mix). The recordist must be aware of these new relationships of dynamic levels, as well as of the possibility that dynamic levels and relationships may function on any hierarchical level of the musical structure.

The recording process places new critical and analytical listening requirements on the recordist, in the area of evaluating loudness. The recordist must be able to focus on changes in dynamics at all levels of perspective, and to quickly switch focus between those levels. The recordist must be able to utilize the skills of listening to loudness relationships, switching between the sound itself, out of context (using critical listening), and the sound and the musical context within which it is found (analytical listening).

Much confusion often accompanies beginning studies in plotting dynamic changes. The listener must always be focused on the process of perceiving and defining changes in loudness levels (which are being applied to music as the artistic element of dynamic levels and relationships).

The listener must be conscious and ever mindful of not confusing other, easily misleading information, as changes in dynamic levels. Some aspects of sound that are often confused for dynamics are distance cues, timbral complexity, performance intensity, sound source pitch register, any information that draws the attention (focus) of the listener (such as a text, sudden entrance of an instrument), environmental cues, and speed of musical information.

It is common for sounds that are most prominent in the listener's focus to not be the loudest sounds in the musical texture. Loudness itself does not create or ensure prominence of the material. The most prominent sound in the listener's attention is often *not* the sound of the highest dynamic level.

This chapter seeks to define actual loudness levels in musical contexts (dynamics), with the exception of the chapter's final section. In that section, the actual loudness of the musical parts as musical balance will be compared against performance intensity (the loudness of the sound sources when they were performed in the recording process).

REFERENCE LEVELS AND THE HIERARCHY OF DYNAMICS

Dynamics have traditionally been described by imprecise terms, such as very loud (fortissimo), soft (piano), medium loud (mezzoforte), and so forth. These terms do not provide adequate information to define the loudness level of the sound source. They merely provide a vocabulary to communicate relative values.

The artistic element of dynamics in a piece of music is judged in relation to context. Dynamic levels are gauged in relation to (1) the overall conceptual dynamic level of the piece of music, (2) the sounds occurring simultaneously with a sound source, and (3) the sounds that immediately precede and follow a particular sound source.

People perceive loudness as relationships between sound sources, in relation to a reference level. Evaluation is more precise, and it is only possible to communicate meaningful information about dynamic levels when a reference level is defined.

An impression of an overall or global intensity level of a piece of music will be the primary reference level for making judgements concerning dynamics. This level is arrived at through a global impression of the intensity level of the performance of the work. It is the perceived performance intensity of the work as a whole that conceptualizes the entire work as a single entity out of time.

Every work can be conceived as having an overall *reference dynamic level (RDL)*. This is the dynamic level that can be assigned to the piece when the listener conceives the form of the work (the piece as existing in an instant). It is the single dynamic level that can represent the overall concept (form) of the piece. This overall dynamic level is a concept of a global impression of the performed dynamic level, or the intensity level of the performance of a piece of music. Performers establish this reference dynamic level in their minds before beginning a performance of any piece of music; recordists must go through a similar process, somewhat in reverse. Recordists establish this level as a reference from which they are able to calculate all other dynamic levels and relationships.

Performance intensity cues are related to timbral changes of the sound sources. Sound sources will exhibit different timbral characteristics when performed at different dynamic levels, with different amounts of physical exertion. The impressions the listener receives, related to the intensity level of the performance (of all the musical parts individually and collectively), will be related to actual dynamic relationships of the musical parts.

The perceived performance intensity in the recording and the perceived dynamic relationships of the musical parts will directly shape the listener's impression of the appropriate loudness level of the recording. The dynamic and performance intensity cues of sound sources that present the primary musical materials (or that are at the center of the listener's focus) will have proportionally more influence than those of sources presenting less significant material.

A single, conceptual dynamic level (at the level of the perceived performance intensity) at which the performance/piece of music is perceived as existing will thus be determined. This is the RDL of the performance (recording/piece of music). It is the reference level that will be used to calculate the dynamic levels of the individual musical parts in relation to the whole, as well as the dynamic contour of the overall program.

Every work will have a specific reference dynamic level (RDL). When a work is divided into separate major sections, and especially separate "movements" (such as a symphony), it is common for each major section to have a different RDL.

The RDL can be perceived as existing anywhere, from "forte" to "piano," and it will be established as a precisely defined dynamic level that will serve as a reference level throughout the work. The listener may transpose the dynamic level relationships, by redefining the reference dynamic level to any other dynamic marking (*mp* redefined to *f*, *pp* into *mf*, etc.). The dynamic levels of the musical materials will then be calculated proportionally to the reference dynamic level. This transposition of the reference dynamic level occurs infrequently for the overall program; it is most common to transpose the dynamic relationships of a single sound source, in relation to the overall ensemble.

The RDL will be used for judging (1) the dynamic relationships of the overall dynamic contour of the program (piece of music), (2) the dynamic levels of the major sections of the work, (3) the dynamic levels of the musical ideas (sound sources) of the work, and (4) the dynamic relationships of the individual dynamic contours of the musical balance of the work.

Timbre changes of performance intensity are important cues in the perception of dynamic levels of acoustic performances. Listeners apply these same cues to recorded sounds, to imagine a live performance, despite the medium. Their reference for performance intensity is their knowledge of the instrument's timbre, as it is played at various levels of physical exertion and with various performance techniques.

Recordists may continue to use the traditional terms for dynamic levels, if they are conscious of a well-defined reference dynamic level. With a defined reference level, the comparative terms can have more significant meaning. The terms will remain imprecise, but they will be more meaningful. Dynamic levels are more precisely defined when sound sources are compared to one another and placed on the appropriate graph, making the use of the traditional terms a mere starting point for evaluating loudness levels.

The terms retain their meanings from musical contexts, where the dynamic terms (such as "mezzoforte") describe a quality of performance and an amount of physical exertion and drama on the part of the performer, as well as being a description of the loudness level. The dynamic terms (such as "forte" (loud) or "piano" (soft)) will always be qualified in relation to an overall impression of the work, which is its reference level.

Even when measuring sound pressure level, a reference level is utilized, to define the decibel. Establishing a reference level is equally important in critical listening situations, as in analytical listening (just discussed) or in measuring SPL. It is possible to acquire the ability to judge relative amounts of differences of loudness levels, by comparing successive or simultaneous sounds, out of musical contexts.

Using one of the sounds as a reference level, the relative amount of loudness above or below can be gauged. Humans cannot accurately remember and transpose precise increments related to the decibel, throughout the hearing range. It is possible to refine loudness level judgements, to gauge distances between the two levels, or even to relate them to the decibel, when the listener has much experience.

To accomplish this listening task, the listener will need to have experience listening for small changes in the particular frequency range of the sound source (at similar sound pressure levels), and will need to have a knowledge of how loudness is perceived differently in different portions of the frequency range and at different sound pressure levels.

In beginning studies, describing the differences in dynamic levels in relative terms (from very slight to very large) will prove most useful. Changing vocabulary to include precise dB increments must involve using test equipment or visual metering, to give the listener experience observing changes of increments related to the decibel, in a wide variety of frequency ranges and sound pressure levels. This is the only way that one can reliably learn what these changes sound like, while equating the sound with the physical characteristics. The skill of judging approximate dB increments, at various SPLs and frequency areas, is often accurately utilized by a wide variety of professionals in the audio industry.

Acquiring this skill will require extended, dedicated listening practice. A suitable exercise for developing this skill can be devised by the reader, to learn to judge differences between sound levels. This should be practiced in the contexts of various sound pressure levels, in various frequency areas, and at various loudness levels (dBs). The exercise will require the listener to learn to recognize the values of a dB, or a multiple thereof, within the context of SPL and frequency area. The listener should move systematically from that which is easily recognized and remembered, gradually through more detailed and diverse materials.

The perceived performance intensity level that serves as our reference for judging the dynamic relationships of the work is utilized at upper levels of the structural hierarchy. The RDL will be used (1) at the highest hierarchical level, to calculate the dynamic contour of the entire program (the complete musical texture), and (2) at the mid-levels of the structural hierarchy, to calculate the dynamic contours of the individual sound sources in musical balance.

At the lower levels of the structural hierarchy, the reference level switches to the global impression of the intensity level of the sound object. The intensity level of the sound source itself is used as a reference, to determine the dynamic contours of the individual sound and its component parts. At these levels of perspective, dynamic contours are plotted of (1) a typical appearance of the overall sound source (the dynamic envelope) and (2) the individual components of the spectrum (spectral envelope). This dynamic contour information seeks to define the sound quality or timbre of the sound source, and will be explored in the next chapter, in that context.

DYNAMIC LEVELS AS RANGES

Dynamic levels are not organized into discrete increments in musical contexts. They are conceived in ranges or areas. Dynamic markings refer to a range of dynamic levels.

The dynamic marking "mezzoforte" does not refer to a precise level, but to a range of dynamic levels between "mezzopiano" and "forte." Many gradations of "mezzoforte" may exist in a certain piece of music. Many instruments can be performing at different levels of loudness, yet be accurately described as being in the "mezzoforte" dynamic range. Entire musical works or performances *may* take place within the range of a certain dynamic marking, yet exhibit striking contrasts of levels.

Figure 10-1 presents the vertical axis that will be used for all graphs plotting dynamics. The dynamic markings are centered within the ranges. A number of sound sources are plotted on the graph, and the reference dynamic level is designated on the vertical axis. Each sound source can be compared to the dynamic levels and contours of the other sources, as well

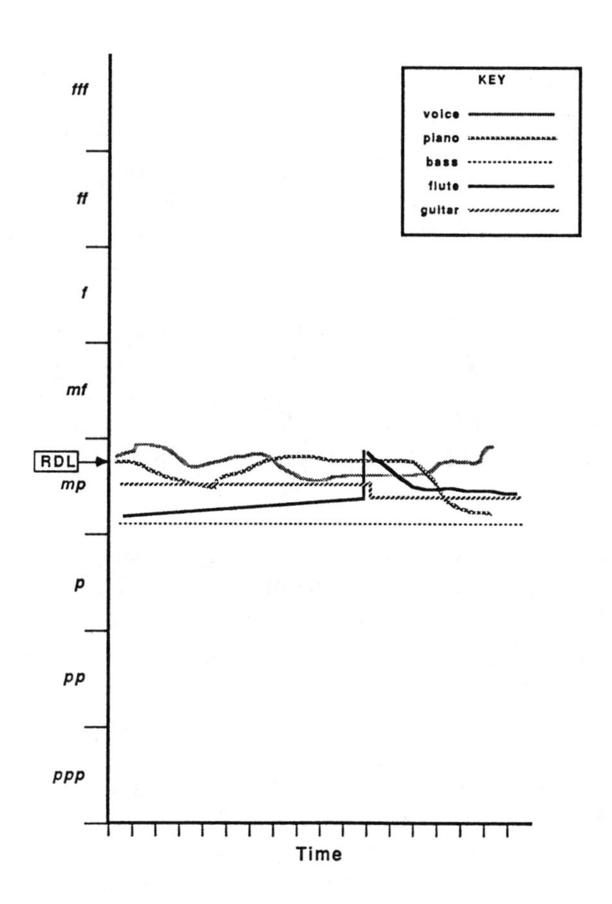

FIGURE 10-1. Dynamic Ranges and Dynamic Contour.

as to the RDL. The unique characteristics of each source are readily apparent from the graph.

Were the description of sound sources limited by traditional dynamic level designations, some sounds would have to be called a "loud mezzopiano," some sounds a "moderate mezzopiano," and others a "soft mezzopiano." The graph, in this instance, circumvents the need for these vague and cumbersome descriptions.

The most extreme boundaries of dynamic ranges, for nearly all musical contexts, will extend from *ppp* to *fff*. Musical examples that contain material beyond these boundaries are rare. The individual graph should incorporate only those ranges that are needed to accurately present the material, and to leave some vertical area of the graph unused above and below the plotted sounds, for clarity of presentation.

DYNAMIC CONTOUR

Changes of the dynamic level, over time, comprise dynamic contour. As noted above, the dynamic levels and relationships occur at all hierarchical levels. The broadest level of perspective will allow the dynamic contour of the overall program to be plotted.

Skill in recognizing the dynamic level of the overall program is developed by plotting *program dynamic contour*. Recordists utilize this skill in many listening evaluations. This high-level graph, or the associated listening skill alone, will be applied to many analytical and critical listening applications.

Program dynamic contour must not be confused with perceived performance intensity, spatial cues, or complexity of textural density. These aspects of recorded sound often present cues that contradict actual dynamic (loudness) level or that alter the perception of the actual dynamic level.

The program dynamic contour graph is created by using the following sequence of events, while listening repeatedly to the musical material.

1. During the first hearing(s), listen to the example, to establish the length of the time line. At the same time, notice prominent instrumentation and activity that will provide cues as to the overall intensity level of the performance.
2. Check the time line for accuracy, and make any alterations. Establish the reference dynamic level (RDL) of the work by conceptualizing the intensity level of the performance, in relation to dynamic level.
3. Notice the activity of the program dynamic contour for boundaries of levels of activity and speed of activity. The boundary of speed will

establish the smallest time unit required to accurately plot the smallest significant change of the element; the boundary of levels of activity will establish the smallest increment of the "Y" axis required to plot the smallest change of the dynamic contour.

4. Begin plotting the dynamic contour on the graph, continually relating the perceived dynamic level to the RDL. First, establish prominent points within the contour; these reference points will be the highest or lowest levels, the beginning and ending levels, points immediately after silences, and other points that stand out from the remainder of the activity. Use the points of reference to judge the activity of the preceding and following material. Focus on the contour, speed, and amounts of level changes, to complete the plotting of the contour. The evaluation is complete when the smallest significant detail has been perceived, understood, and added to the graph.

The information is plotted on a program dynamic contour graph. The graph incorporates the following:

1. The dynamic area designations for the "Y" axis, distributed to complement the characteristics of the musical example.
2. The reference dynamic level is designated as a precise level on the "Y" axis.
3. The "X" axis of the graph is dedicated to a time line that is devised to follow an appropriate increment of the metric grid or of real time (depending on the material and context).
4. A single line is plotted against the two axes (the dynamic contour of the composite program is the shape of this line).

The program dynamic contour of Elton John's "Sorry Seems To Be The Hardest Word" is plotted in Figure 10-2. The program dynamic contour graph shows the changes in overall dynamic level throughout the work.

MUSICAL BALANCE

Plotting all the sound sources in a musical texture by their dynamic contours provides the *musical balance graph.* The graph will show the actual dynamic level of the sound sources, in relation to the RDL, as established earlier. The graph will dedicate a source line to each sound source and will plot the dynamic contours of each source against a common time line. This graph is a representation of the mix of the work.

The musical balance graph will not incorporate performance intensity,

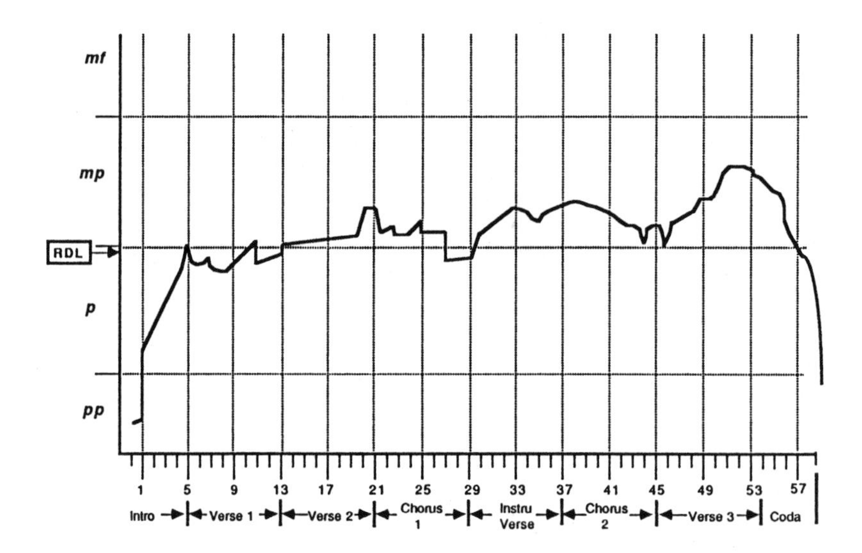

FIGURE 10-2. Program Dynamic Contour—Elton John: "Sorry Seems to Be The Hardest Word."

sound quality, or distance cues. These cues are often confused with perceived dynamic level and are purposefully avoided in this exercise. This graph is solely dedicated to appreciating the dynamic levels, contours, and relationships of the sound sources. The exercise should follow the sequence:

1. During the first hearing(s), listen to the example to establish the length of the time line. At the same time, notice prominent instrumentation and the activity of their dynamic levels against the time line.
2. Check the time line for accuracy, and make any alterations. Establish a complete list of sound sources (instruments and voices), and sketch the presence of the sound sources against the completed time line.
3. Determine the reference dynamic level by listening to the performance intensities of the sound sources.
4. Notice the activity of the dynamic levels of the sound sources (instruments and voices), for boundaries of levels of activity and speed of activity. The boundary of speed will establish the smallest time unit required to accurately plot the smallest significant change of dynamic level; the boundary of levels of activity will establish the smallest increment of the "Y" axis required to plot the smallest change of dynamics. This step will establish the perspective of the graph.
5. Begin plotting the dynamic contours of each instrument or voice on the

graph. Keeping the RDL clearly in mind, establish the beginning dynamic levels of each sound source. Next, determine other prominent points of reference. Use the points of reference to judge the activity of the preceding and following material. Focus on the contour, speed, and amounts of level changes, to complete the plotting of the dynamic contours.

The listener should periodically shift focus, to compare the dynamic levels of the sound sources to one another. This will aid in developing the dynamic contours, and will keep the listener focused on the relationships of dynamic levels of the various sources. The evaluation is complete when the smallest significant detail has been incorporated into the graph.

The musical balance graph incorporates the following:

1. The dynamic area designations for the "Y" axis, distributed to complement the characteristics of the musical example.
2. The reference dynamic level is designated as a precise level on the "Y" axis.
3. The "X" axis of the graph is dedicated to a time line that is devised to follow an appropriate increment of the metric grid.
4. A single line is plotted against the two axes for each sound source.
5. A key will present the names of the sound sources to the related color, number, or composition of source lines on the graph.

Figure 10-3 presents a musical balance graph of The Beatles' "Lucy In The Sky With Diamonds."

PERFORMANCE INTENSITY VERSUS ACTUAL DYNAMIC LEVEL

Performance intensity is the dynamic level at which the sound source was performing when its was recorded. In many music productions, this dynamic level will be altered in the mixing process of the recording. The performance intensity of the sound source and the actual dynamic level of the sound source in the recording will most often send conflicting information to the listener.

The dynamic levels of the various sound sources of a recording will often be at relationships that contradict reality. Sounds of low performance intensity often appear at higher dynamic levels in recordings than sounds that were originally recorded at high performance intensity levels. Important information can be determined by plotting performance intensity against musical balance, for some or all of the sound sources of a recording. This will often provide significant information on the relationships of sound

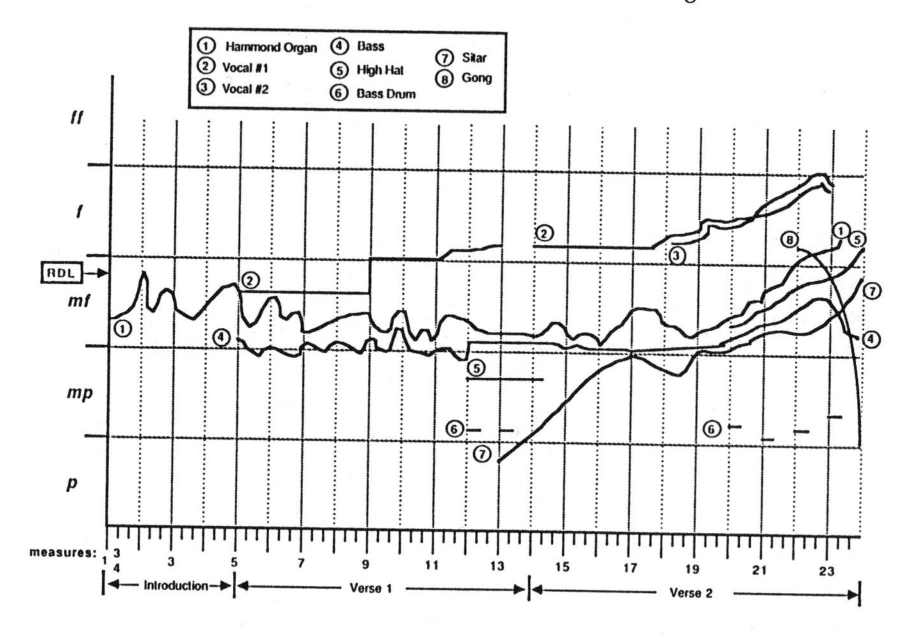

FIGURE 10-3. Musical Balance Graph—The Beatles: "Lucy In The Sky With Diamonds."

sources, the overall dynamic and intensity levels of the work, and the mixing and compositional processes.

Performance intensity is plotted as the dynamic levels of the sound sources, as the listener perceives the intensity of the original performance. The listener will judge the intensity of the original performance, through timbre cues. The listener will make use of prior knowledge of the sound qualities of instruments and voices, as they exist when performed at various dynamic levels, to make these judgements. The reference for performance intensity is the listener's knowledge of the particular instrument's timbre, as that instrument is played at various levels of physical exertion and with various performance techniques. A reference dynamic level is not applicable to this element.

Musical balance is plotted as the dynamic levels of the sound sources, as the listener perceives their actual loudness levels in the recording itself, as was discussed earlier.

The performance intensity/musical balance graph incorporates the following:

1. The dynamic area designations in two tiers for the "Y" axis, distributed to complement the characteristics of the musical example. One tier is

dedicated to musical balance, and one tier is dedicated to performance intensity.

2. The reference dynamic level of the musical balance tier is designated as a precise level on the "Y" axis. An RDL is not relevant to the performance intensity tier.
3. The "X" axis of the graph is dedicated to a time line that is devised to follow an appropriate increment of the metric grid.
4. A single line is plotted against the two axes for each sound source, on each tier of the graph. Each sound source will appear on both tiers. The same composition or color line is used for the source on each tier of the graph.
5. A key is required to clearly relate the sound sources to their respective source line. The same key applies to both tiers of the graph.

The musical balance sequence described above applies here as well. The sequence is directly adapted for the performance intensity tier. The same steps are applied to the perceived intensity of the performance that were previously applied to the perceived, actual loudness level.

Figure 10-4 will allow the reader to observe some of the differences

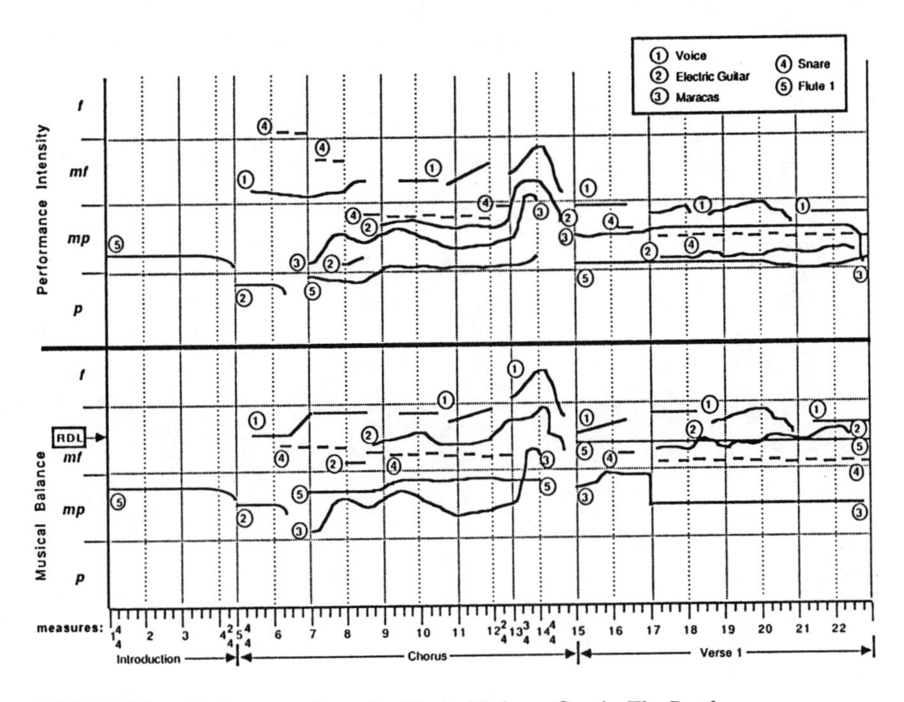

FIGURE 10-4. Performance Intensity/Musical Balance Graph—The Beatles: "Strawberry Fields Forever."

between actual loudness levels and the recorded performance intensities in The Beatles' "Strawberry Fields Forever." As a exercise, the reader should determine the musical balance and performance intensities of the other sound sources.

The musical balance and the performance intensity graphs will function at the same level of perspective as the textural density graph. When evaluated jointly, these three artistic elements will allow the listener to extract much pertinent information about the mixing and recording processes, as well as the creative concepts of the music.

11

The Evaluation of Sound Quality

People in the audio industry, as in all industries, work together towards common goals. In order to achieve these goals, there must be clear and effective communication. A vocabulary for communicating specific, pertinent information about sound quality does not currently exist. This chapter will present a process for evaluating sound quality that can also be applied to defining the characteristics of timbre. The process is projected towards establishing a vocabulary for meaningful communication.

The process must be easily adapted to any sound evaluation. It must be able to describe sound quality (accounting for the contexts within which sounds exist), and it must be usable for evaluating a sound out of context (as an abstract, sound object).

The critical listening process and the technical needs of the audio industry are often juxtaposed with creative applications and analytical listening processes. These differences are articulated, as are conceptions of sound quality that can occur at various hierarchical levels.

While sound quality is the perception of sound for its global form or shape, this conception of sound exists at a number of levels in human perception. These concepts are central to an evaluation, as they allow for the realization of sound as an "object" (available for evaluation out of time), at all levels of detail in the listener's perception.

COMMUNICATING ABOUT SOUND QUALITY

Communicating information about sound quality is central to all facets of music production. Nearly all positions in the audio industry need to communicate "about" the content or quality of sound. Yet a vocabulary for describing sound quality or a process for objectively evaluating the components of sound quality do not exist.

Describing the characteristics of sound quality through associations with the other senses (through terminology such as "dark," "crisp," or "bright" sounds) is of little use in communicating precise and meaningful information "about" the sound source. "Bright" to one person may be associated with a narrow, prominent area of spectral activity around 15 kHz, throughout the sound source's duration; to another person, the term may be associated with fast transient response in a broader frequency area around 8 kHz, present only for the initial third of the sound's duration; a third person might easily provide a different, yet equally acceptable, definition of "bright," within the context of the same sound. The three people would be using different criteria of evaluation, and would be identifying markedly different characteristics of the sound source, yet they would be calling three potentially quite different sounds the same thing: "bright." This terminology will not communicate specific information about the sound. This type of terminology will not be universally understood; it will not have the same meaning to all people.

Analogies such as "metallic," "violin-like," "buzzing," or "percussive" might appear to supply more useful information about the sound than the inter-sensory approach, described in the last paragraph. This is not so. Analogies are, by nature, imprecise; they compare a given sound quality to a sound that the individuals already "know." A common reference between the individuals attempting to communicate is often absent. Sounds have many possible states of sound quality.

"Violin-like" to one person may actually be quite different to another person. One person's reference experience of a "violin" sound may be an historic instrument built by Stradivari and performed by a leading artist at Carnegie Hall. Another person may use the sound of a bluegrass fiddler, performing a locally-crafted instrument in the open air, as his or her reference for defining the sound quality of a "violin." The sound references are equally valid for the individuals involved, but the references are far from consistent and will not generate common ground for communication. The sound qualities of the two sounds are strikingly different; the two people will be referencing different sound characteristics, while using the same term. An accurate exchange of information will not occur.

The imprecise terminology related to sound quality is at its most extreme when sounds are categorized by mood connotations. Sound qualities are sometimes described in relation to the emotive response they invoke in the listener. The communication of sound quality through terminology such as "somber," for example, will mean very different things to different people. Such terminology is so imprecise that it is useless in communicating meaningful information about sound.

People can only communicate effectively by using common experiences or knowledge. The sound source itself is presently the only common experience between two or more humans.

As people hear sounds, they make many individualized interpretations and personal experiences. These individual interpretations and impressions are present within the human perceptual functions of hearing and evaluating sound; they cause individualized distortions of the meaning and content of the sound. Therefore, people's interpretations and impressions are of little use in communicating about sound. Humans have few listening experiences that are common between individuals and that are available to function as the reference necessary for a meaningful exchange of information (communication). This absence of reference experiences and knowledge makes it necessary for the sound source itself to be described. Meaningful communication about the sound quality of the sound source will not be precise and relevant, without such a description.

The states and activities of the physical characteristics of the sound will be described, in this book's communications about sound quality. This approach to evaluating sound quality will require a knowledge of the components of timbre and their functions in creating sound quality. Meaningful communication between individuals is possible when the actual sound source is described through defining the activities of its component parts.

By describing the states and activities of the physical components of a sound, people may communicate precise, detailed, and meaningful information about the sound source. The information must be communicated clearly and objectively. All the listener's subjective impressions about the sound, and all subjective descriptions in relation to comparing the sound to other sounds, must be avoided, for meaningful communication to occur. Subjective information does not transfer to another individual. As people attempt to exchange their unique, personal impressions, the lack of a common reference does not allow for the ideas to be accurately exchanged.

Meaningful communication about sound quality can be accomplished by describing the values and activities of the physical states of the components of timbre. Sounds will be described by the characteristics that make them unique. These characteristics are the activities and states that occur in the component parts of the sound source's timbre. The characteristics have been reduced to: definition of fundamental frequency, dynamic envelope, spectrum (spectral content), and spectral envelope.

Meaningful information about sound quality can be communicated by verbally describing the values and activities of the physical states of characteristics of timbre in a general way. Information is communicated in a more detailed and precise manner, by graphing the activity.

SOUND QUALITY IN CRITICAL LISTENING AND MUSICAL CONTEXTS

Describing sound quality in and out of musical contexts are skills that are important for audio professionals. In both contexts, sound quality evaluations occur at all levels of perceptual hierarchies, as well as at all perspectives.

Outside musical contexts, the audio professional is concerned with sound quality as the integrity of the audio signal. Through critical listening skills, sound quality of the audio program is evaluated. The evaluation is in relation to the activities and states of the components of sound that occur in the program. Critical listening evaluation is often concerned with the technical quality, or integrity, of the recording.

The technical quality of recordings is the amount of wanted sound versus unwanted sound, as well as the amount of distorted sound within a recording. It should be the goal of all recordings to be of the highest technical quality and to be void of all distortion and all unwanted sound. The technical quality of recordings is defined by the types of unwanted sounds and sound events, and by numerous types of distortions in the dimensions of frequency/pitch and amplitude/loudness, caused by the recording and reproduction equipment and processes.

The focus of the listener may be at any level of listening perspective, with the listener able to quickly and accurately shift perspectives. The listener may be evaluating the technical quality of the sound, for information related to frequency response (or spectral content, or spectral envelope), or the listener may be listening for transient response (or dynamic envelope, or dynamic contour of a specific frequency area). As examples of extremes of perspective: the listener may focus on the sound quality of the overall program, or the listener may be listening at the close perspective of focusing on the sound quality of a particular characteristic of a single, isolated sound source.

Critical listening involves evaluating sound quality, to define what is physically present, to identify the characteristic qualities of the sound being evaluated, or to identify any undesirable sounds or aspects that influence the integrity of the sound (audio signal). This process and conceptualization is performed without considering the function and meaning of the sound and without taking into account the context of the sound.

Analytical listening involves evaluating sound quality, to identify its characteristic qualities in relation to the context of the sound. Analytical listening will seek to define the sound quality in terms of what is physically present, but will then relate that information to the musical context in which the sound material is presented and perceived.

Textural density evaluates pitch-related characteristics of sound quality, at a high level in the structural hierarchy. This evaluation may or may not occur in a musical context. Pitch area analysis (such as the ones performed on percussion sounds in Chapter 9) is another evaluation of pitch-related characteristics of sound quality, only at a lower hierarchical level. It also does not necessarily take place in relation to the musical context. The textural density analysis that took place in Chapter 9 was in a musical context; the pitch area analysis of the tom drum was out of musical context.

Both of these studies are simple components of timbre, or sound quality, analysis. They both define the pitch-component information that leads to defining timbre, or evaluating sound quality. This process is related to the evaluation of the spectral content of a sound.

In the same way, dynamic contour analyses are related to sound quality analysis, at various structural levels.

SOUND QUALITY AND PERSPECTIVE

Sound quality is the perception of sound as a single concept or entity. Sound is conceived in its global form or shape, at a number of levels of perceptual hierarchies. People recognize sound quality at all levels of detail in their perception. This allows for understanding sound quality as an "object" (available for evaluation out of time) or in relation to its musical context (as a sound event), at all levels of perceptual detail.

People are able to recognize sound quality as the global qualities of overall program (or the entire musical texture), groupings of similar or similarly acting sounds within the overall program (such as a brass section within an orchestra, or the rhythm section of a jazz ensemble), overall impressions of individual sound sources (instruments, voices, synthesizer patches, special effects), or specific sounds generated by individual sound sources (individual expressive vocal sounds, a specified voicing of a guitar chord, a single sound's timbre characteristics). Listeners even recognize the sound quality of acoustic spaces, as environmental characteristics (next chapter).

At these very different levels of the perceptual hierarchy, listeners recognize sound quality as the concept that makes a sound a single, unique entity—an overall form of the sound. This global quality is evaluated to determine specific information, into the aspects that make all sounds unique.

This evaluation may take place out of musical contexts. In these critical listening applications, evaluation is performed out of the time line of musical context. Instead, clock time is used to evaluate the changes that have occurred over time. While listeners are conceiving sound quality as the shape of the sound "in an instant, or out of time," sound only exists in time. Sound can only be accurately evaluated as changes in states or values of the component parts that occur over time. Sound quality in critical listening applications approaches the sound as an isolated, abstract object.

The evaluation may also take place within musical contexts. In these instances, the time of the metric grid will be used, if it is present. Evaluating sound quality will be focused on the musical relationships of the material. The textural density analysis of Chapter 9 is a suitable example of the evaluating the pitch aspects of sound quality, in relation to the musical material. If the textural density graph was coupled with a musical balance graph, much information on the sound quality of the entire program (in a musical context) would be available to the listener. Sound quality in analytical listening applications approaches the sound for its relationship to other sounds, to the musical texture as a whole, and to the musical message of the work.

The sound quality of the entire program, or of the individual sound sources, may be supplying the most significant musical information in certain pieces of music. This concept of music composition (that can be explained through equivalence) is quite prominent in many very different styles of music. Throughout the twentieth century, a type of writing—*sound mass composition*—has evolved through the work of composers Edgar Varèse, George Antheil, Krzysztof Penderecki, and Yannis Xenakis (to name only a few). This music places an emphasis on the dimensions of the overall musical texture (or sound mass), or is based on sound quality relationships within the overall musical texture.

The concept of giving musical significance to the sound quality of the entire program, to the relationships of sound qualities, and to textural density can be found in a wide variety of popular works from the past 30 years. Many examples of these ideas exist, although this concept is used in isolated areas in most works. Two notable works that should be consulted by the reader are The Beatles' "A Day In The Life" and U2's "Where the Streets Have No Name." The two groups use the concept of textural density in strikingly different ways: the two textures have very different sound properties, and the two works present the sound mass ideas as very different sections in the works (U2 as an introduction, The Beatles as a bridge). Textural density is the primary musical element in those sections of the works, and it functions differently in each work.

Sound quality will be evaluated in relation to spectral content, spectral envelope, and dynamic contour, in both musical contexts and critical listening evaluations. Sound quality evaluation will be approached in this way, at all levels of the perceptual hierarchy. At the highest levels, individual sound sources that make up the whole program might be conceived as individual spectral components. At the lowest level, individual spectral components are evaluated, and individual evaluations may be performed for each occurrence of a sound source.

EVALUATING SOUND QUALITY

Individual sound sources may be analyzed for their contributions to the sound quality of the overall program. In musical contexts, this evaluation will compare the sound sources to the overall sound quality, through their individual dynamic contours (creating musical balance), their pitch area (creating textural density evaluations), and their spatial characteristics (Chapter 12). In critical listening, the contributions of the individual sound sources to the overall program will be approached in relation to the same dimensions, but without relation to musical time or context.

Individual sound sources may also be evaluated for their unique sound quality. In this way, individual sound sources are evaluated as sound objects. The individual sounds are evaluated, to define their unique characteristics, through a detailed examination of the states and activities of their component parts, out of the musical context. This is the most widely applied use of sound quality evaluation, being used from signal processing to evaluating the performance of audio devices. This approach is also used to define the general characteristics of a sound source, such as the sound quality of a guitar part in a recording, or it may take the form of a detailed evaluation of a particular guitar sound.

Often, a sound quality evaluation will be performed on a single, isolated presentation of the sound source (presentation being pitch level, performed dynamic level, method and intensity of articulation, etc.). This allows for meaningful comparison between different performances of the same source, or of a different sound source performing similar material. Evaluations of sound quality will seek to define the states and activities of the sound source's (1) dynamic envelope, (2) spectral content, and (3) spectral envelope, as well as the listener's carefully evaluated perception of (4) pitch definition.

Sound quality is defined by the states and values of the physical dimensions of the sound source (1, 2, and 3). The listener's perception of the definition of fundamental frequency (4) will help define the loudness level of the fundamental frequency, in relation to the remainder of the sound's spectrum. This may be important in defining the sound quality. The four components of sound quality are evaluated throughout the duration of the sound material, and are plotted against a single time line.

The reader will first define the time line of the sound. A clock, or stopwatch, should be used as a reference. Determining increments within the time line, as well as suitable reference points, requires greater skill and will require a number of listenings.

The dynamic contour of the sound, or the sound's overall dynamic level, as it changes throughout its duration, is reasonably apparent at first listenings. Difficulties may arise with confusing loudness changes, with spectral complexity changes; the two may be related, but they may be very different.

An RDL will be required for mapping the dynamic contour. The RDL will be determined by the intensity level at which the source was performed. The intensity level itself is the RDL, but it may be transferred to a dynamic level of "mezzoforte" or "forte" if the RDL is at a level that would make dynamic contours difficult to plot (i.e., the actual RDL being *pp*). If the RDL is transferred to a different dynamic level, this transfer must be noted as part of the evaluation. The same reference dynamic level will be used in the spectral envelope tier, explained in the following paragraphs. In this way, the same reference level functions on two levels of perspective, just as the same reference dynamic level functions for both program dynamic contour and musical balance.

The reader may hear few or no spectral components at the beginning of his or her studies. Harmonics and overtones fuse to the fundamental frequency, and people have been conditioned to perceive all this information as part of a whole (the global sound quality). To a great extent, evaluating sound quality works against all our learned listening techniques and our previous listening experiences. Much patience will be required. Practice and repetitive listening must be undertaken, to acquire the skills of accurately recognizing spectral components and of accurately tracking dynamic contours of the components that make up spectral content.

The reader may wish to use a tone generator, to assist in identifying the frequency level (to be translated into pitch level) of prominent harmonics and overtones. This will prove helpful in initial studies. A steep, tunable filter may also be of use, in determining the pitch aspects of the spectrum.

The dynamic contours of the individual components of the spectrum will be mapped against the time line. The individual dynamic contours will be calculated against the reference dynamic level that was identified for the sound's overall dynamic contour, above. All the spectral components that are identified in the previous paragraphs should be present on the spectral envelope tier, including the fundamental frequency.

The definition of the fundamental frequency is often somewhat stable within a given sound. Changes in pitch quality are most commonly found between the onset and the body of the sound. Pitch quality is often placed on a continuum between the two boundaries of well-defined pitch, or is "precisely pitched" (as a sine wave) though completely void of pitch or "non-pitched" (as white noise). Often, the definition of fundamental frequency can be verbally described as having a certain pitch quality for a certain portion of its duration and then another certain quality for the remainder of its duration. The "pitch definition" tier of the sound quality characteristics graph is not always required, but this aspect of the sound must always be addressed.

The process of evaluating sound quality should follow this sequence of events:

1. During the first hearing(s), listen to the example, to establish the length of the time line. At the same time, notice prominent states and activity of the components (especially dynamic contour and spectral content) against the time line.
2. Check the time line for accuracy, and make any alterations.
3. Notice the activity of the component parts of sound quality, for their boundaries of levels of activity and speed of activity.

   The speed boundaries will establish the smallest time unit required in the graph to accurately present the smallest significant change of the element. The boundary of levels of activity will establish the smallest increment of the "Y" axis required to show the smallest change of each component (dynamic contour, spectral content, spectral envelope).
4. Begin plotting the activity of the dynamic envelope on the graph. First, establish the RDL, then establish the beginning and ending dynamic levels of the sound. The highest or lowest dynamic levels are the next to be determined; place them against the time line. Use these levels as points of reference, to judge the activity of the preceding and following material; alternate focus on the contour, speed, and amounts of level changes, to complete the plotting of the dynamic contour. The evaluation is complete when the smallest significant detail has been perceived, understood, and added to the graph. The smallest time increment of the time line may need to be altered at this stage, to allow the dynamic contour to be clearly presented on the graph.
5. Plot the states and activity of the spectral content on the graph. Often, spectral components remain at the same level (or state) throughout the sound source; certain instruments have prominent formants and overtones that change in pitch level (exhibit activity) over the duration of the sound. First, establish the frequency/pitch levels of the prominent spectral components and the fundamental frequency. Map the presence of these frequencies against the time line, and map any changes in levels of these pitches/frequencies against the time line. Certain spectral components may not be present throughout the duration of the sound; it is not unusual for harmonics and overtones to enter and exit the spectrum. The evaluation is complete when the all the spectral components that can be perceived by the listener are added to the graph.

   Accuracy and detail will increase markedly with experience and practice on the part of the listener. With time and acquired skill, this process will yield much significant information on the sound. Initial attempts may not yield enough information to accurately define the sound source, but will improve with time and effort.

6. Begin plotting the activity of the spectral envelope on the graph. First, establish the beginning and ending levels of each of the spectral components that were identified in 5. For each of the spectral components, in logical sequence, determine the highest or lowest dynamic levels and any other prominent points of reference within the dynamic contours. Use these points of reference to evaluate the preceding and following material; alternate focus on the contour, speed, and amounts of level changes, to complete the plotting of the activity. The dynamic envelopes of all the spectral components are plotted on this tier. The evaluation is complete when the smallest significant dynamic level change has been incorporated into the graph.
7. Pitch definition calculations are made as a final step in sound quality evaluations of sound sources. The prominence of the fundamental frequency and the number and relative loudness levels of harmonics will do much to dictate the pitch quality of the sound source.

Many listenings will be involved for each of the previous steps. Each listening should seek specific new information and should confirm what has already been noticed about the material. Before listening to the material, the listener must be prepared to extract certain information, to confirm previous observations, and to be receptive to new discoveries about the sound quality (Fig. 11-1). The listener should check previous observations often, although his or her listening attention may be seeking new information.

The information is plotted on a *sound quality characteristics graph*. The graph incorporates the following:

1. A multitiered "Y" axis, distributed to complement the characteristics of the musical example: one tier with dynamic areas for dynamic envelope (with notated RDL), one tier with pitch area registral designations for spectral content, one tier with dynamic areas for spectral envelope (with notated RDL), and perhaps a fourth tier designating a pitched to non-pitched continuum.
2. The "X" axis of the graph is dedicated to a time line that is devised to follow an appropriate increment of clock time (exact increment will vary depending on the material).
3. Each spectral component is plotted as a single line, against the two axes,

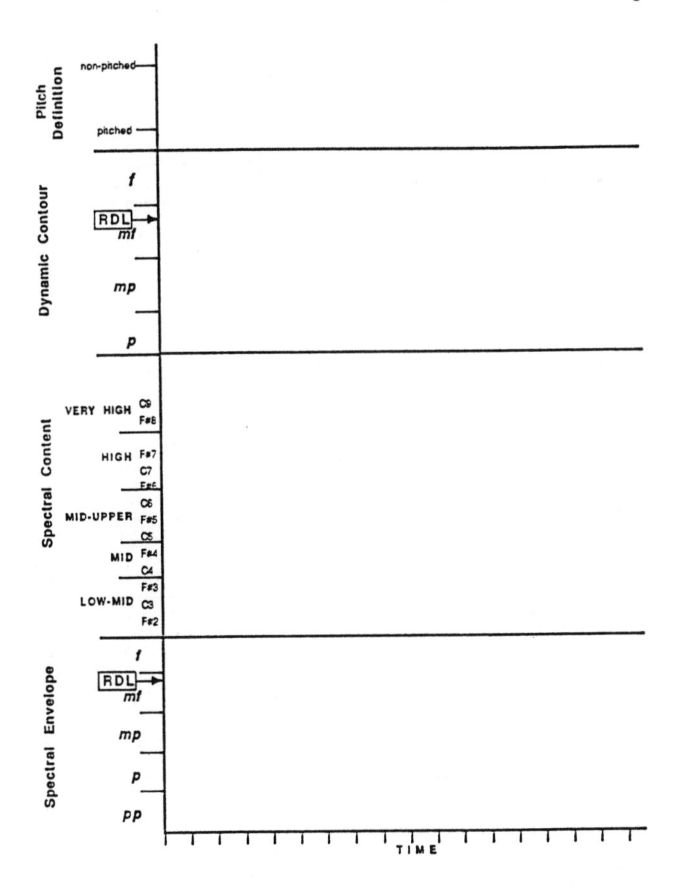

FIGURE 11-1. Sound Quality Characteristics Graph.

its pitch characteristics on the spectral content tier and its dynamic contour on the spectral envelope tier.

4. Each spectral component's line should have a different color or composition, to allow the viewer of the graph to compare the two tiers.

Figures 11-2 and 11-3 provide sound quality evaluations of the first four synthesized sounds from Peter Gabriel's "Mercy Street." The sound sources are numbered in the order of their entrance, at the beginning of the piece. The first appearances of the sound sources are not necessarily those that were evaluated. Appearances of the sound sources, where their characteristics are not being masked by other sound sources or by performance technique, and appearances at similar pitch levels

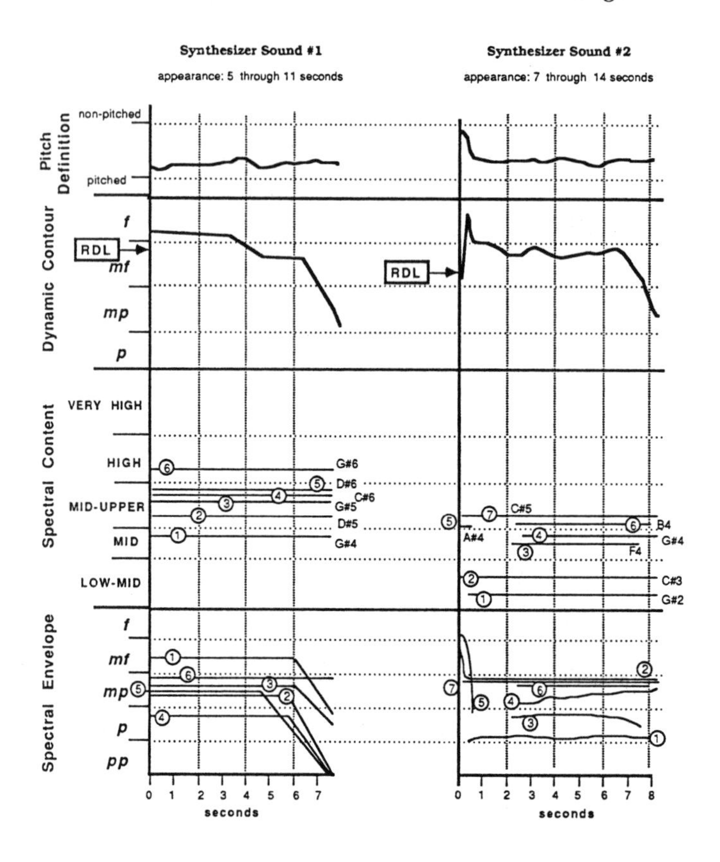

FIGURE 11-2. Sound Quality Characteristics Graphs—Peter Gabriel: "Mercy Street"; Synthesizer Sounds #1 and #2.

were used for evaluating their sound quality. As an exercise, add detail to these four sound quality evaluations, while checking the existing information for accuracy.

The ability to evaluate and communicate about sound quality and timbre is extremely important for a synthesist. It is also required of nearly all positions in the audio industry. The sound quality evaluations allow the reader to quickly recognize the unique characters of the plotted sounds. The skills gained through the previous chapters are brought together in the many steps of sound quality (and timbre) evaluation. Recognizing and understanding the characteristics of sound quality are the first steps towards communicating accurate and relevant information about sound.

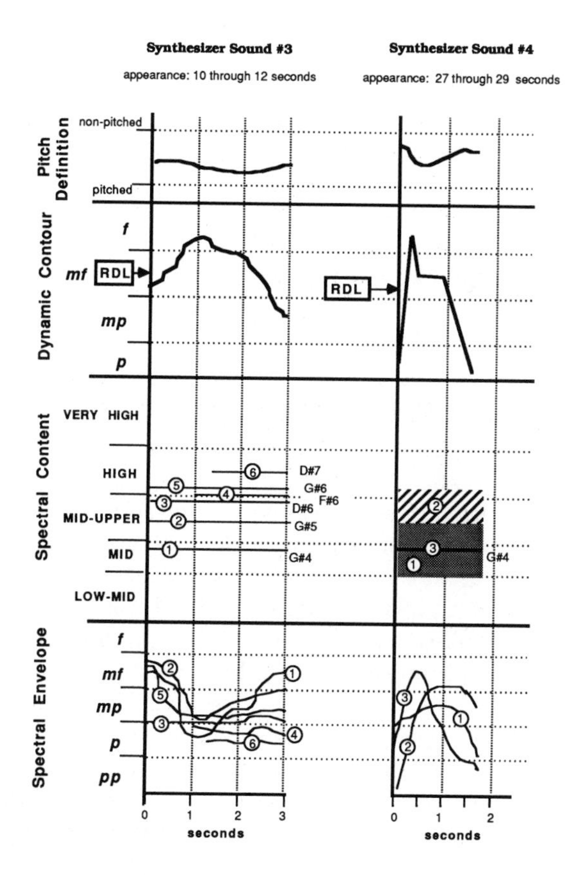

FIGURE 11-3. Sound Quality Characteristics Graphs—Peter Gabriel: "Mercy Street"; Synthesizer Sounds #3 and #4.

12

Evaluating the Spatial Elements of Reproduced Sound

The spatial characteristics and relationships of sound sources are an integral part of music and audio productions. Spatial elements are precisely controllable in audio recording, and a sophisticated usage of these elements has developed in audio and music productions. Spatial relationships and characteristics contribute significant information in current music productions, and are of significant importance in critical listening applications.

Evaluating the spatial characteristics of recordings covers three primary areas: localization on a single horizontal plane in front of the listener, localization in distance from the listener, and recognizing environmental characteristics. The elements of environmental characteristics and distance illusions further interact and create other sound characteristics that must be evaluated.

The recordist must be able to evaluate sound in relation to these characteristics, to properly evaluate recorded/reproduced sound. The skills required to evaluate the spatial characteristics of a recording have been gradually developed throughout the previous four chapters. The skills of sound quality evaluation, time judgements, pitch estimation, and dynamic contour mapping will all be further utilized (from a new perspective), in recognizing and evaluating the spatial elements of reproduced sound. Developing these skills will require much patience and practice.

Many of the concepts of the spatial elements have not previously been well defined. The length of this chapter is the result of the number of important spatial elements of sounds in recordings, the methods one must utilize to perform meaningful evaluations of these elements, and the explanations required of new concepts.

Spatial characteristics and relationships are used as artistic elements in music productions. They are used as primary and secondary elements that help to shape the unique character of musical ideas. Space has the potential of being the most important artistic element in a musical idea, but most often serves to support other elements. It may support other elements by delineating musical materials, by adding new dimensions to the unique character of the sound source or musical idea, and/or by adding to the motion or direction of a musical idea.

The spatial relationships of reproduced sounds are conceived by the listener, through the concepts of sound stage and imaging. The listener will imagine a performance space, wherein the reproduced sound can exist during the re-performance of listening to the recording. The listener will conceptually place the individual sound sources at specific locations within this perceived performance environment.

The recording represents an illusion of a live performance. The listener will conceive the performance as existing in a real, physical space, because the human mind will conceive of any human activity in relationship to the known, physical experiences of the individual. The recording will appear to be contained within a single, perceived physical space (the performance environment), because in human experience people can be in only one place at one time.

The perceived performance environment will have audible characteristics (a sound quality) that are established in two ways. The characteristics of the perceived performance environment may be established by a set of environmental characteristics applied to the overall program (applying an environment to the final mix). Most often, the perceived performance environment is a composite of many perceived environments and environmental cues. In these instances, the perceived performance environment is conceived by the listener, through environmental characteristics that are (1) common or complementary between the environments of the individual sound sources, (2) prominent characteristics of the environments of prominent sound sources (source that presents the most important musical materials, or the loudest, nearest, or furthest sound sources, as examples), and/or (3) the result of environmental characteristics found in both of these areas.

Further, listeners will perceive themselves to be placed at a specific location within the perceived performance environment. They will be especially aware of their relationships to the side walls and any objects (balconies, seating, etc.) in the performance environment, to the wall behind and the ceiling above their location, and to the front wall of the performance environment. The listeners will also calculate their location with respect to the front of the sound stage.

Within this perceived performance environment is a two-dimensional area (horizontal plane and distance), where the performance is occurring: the sound stage (Fig. 12-1). The sound stage is the location within the

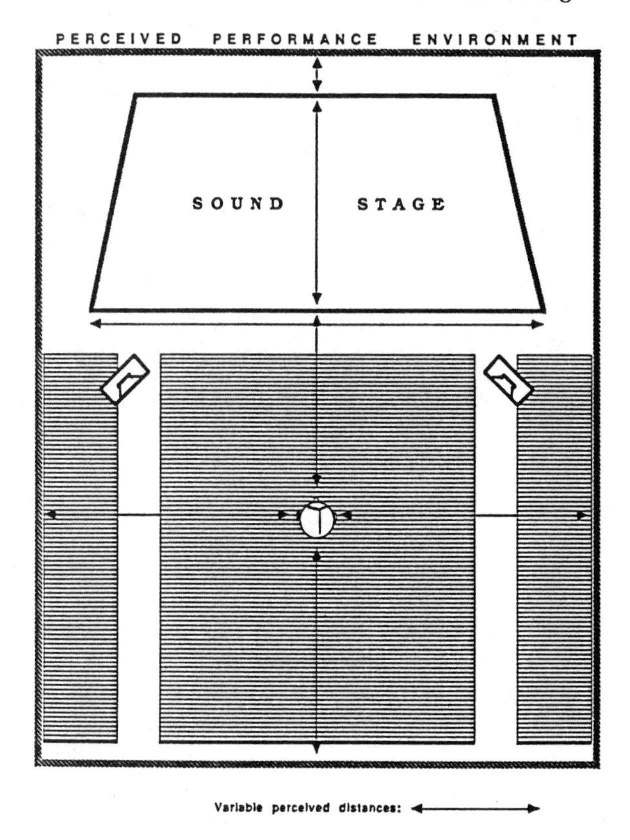

FIGURE 12-1. The Sound Stage Within the Perceived Performance Environment.

perceived performance environment, where the sound sources are perceived to be collectively located, as a single ensemble.

The area of the sound stage may be any size. The size of the sound stage may appear to be anything from the size of a pinhead (an infinitesimally small world) to occupying an area extending from immediately in front of the listener to a location (spanning a distance) well beyond the listener's reality or imagination (perhaps conceived as being an area beyond the size of anything known within human experience) and, within the horizontal plane, filling an area beyond the stereo array.

The sound stage may be located at any distance from the listener. The placement of the front edge of the sound stage may be immediately in front of the listener or at any conceivable distance from the listener.

The depth of the sound stage is determined by the perceived distances of the sound sources from the listener. The sound source that is perceived as being nearest to the listener will mark the front edge of the sound stage. The sound source that is perceived as being furthest from the listener will mark the end of the sound stage and will also help to establish the rear wall of the perceived performance environment. The perceived location of the rear boundary (wall) will be determined by the relationship of the furthest sound source to its host environment. The rear wall of the perceived performance environment may be located immediately behind the furthest sound source, or some space may exist between the furthest sound source and the rear wall of the sound stage/perceived performance environment.

All sound sources will occupy their own location in the sound stage. Two sound sources cannot be conceived as occupying the same physical location; human sensibilities will not allow this to occur. It is possible for different sound sources to occupy significantly different locations within the sound stage, anywhere between the two boundaries. Imaging is the perceived location of the individual sound sources within the two perceived dimensions of the sound stage (Fig. 12-2). Sources are located within the sound stage by their angle (on the horizontal plane) and distance from the listener.

The imaging of sound sources will be influenced by the characteristics

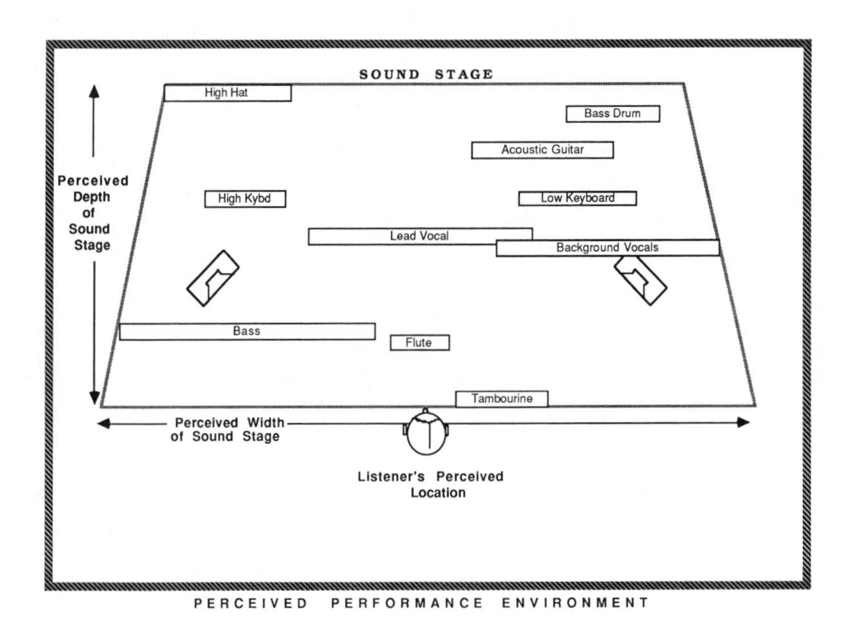

FIGURE 12-2. Sound Source Imaging Within the Sound Stage.

of the unique performance environments of the individual sound source, as well as their placement within their own environment. In current music productions, it is common for each instrument (sound source) to be placed in a different host environment. The host environment of the individual sound source (a perceived physical space) is imagined to be existing within the perceived performance environment (space). This creates an illusion of a *space* existing *within* another *space.*

The environments of the sound sources and the overall program may be any size. The acoustical characteristics of any space may be simulated by modern technology. The sound sources may be processed so that the cues of any acoustical environment may be added to the individual sound source, to any group of sound sources, or to the entire program. Not only is it possible to simulate the acoustical characteristics of known, physical spaces, it is possible to devise environment programs that simulate open air environments (under any variety of conditions) and programs that provide cues that are acoustically impossible within the known world of physical realities.

Figure 12-3 presents an easily accomplished set of environmental relationships, with individual sound sources appearing to be performed in very different and unique environments:

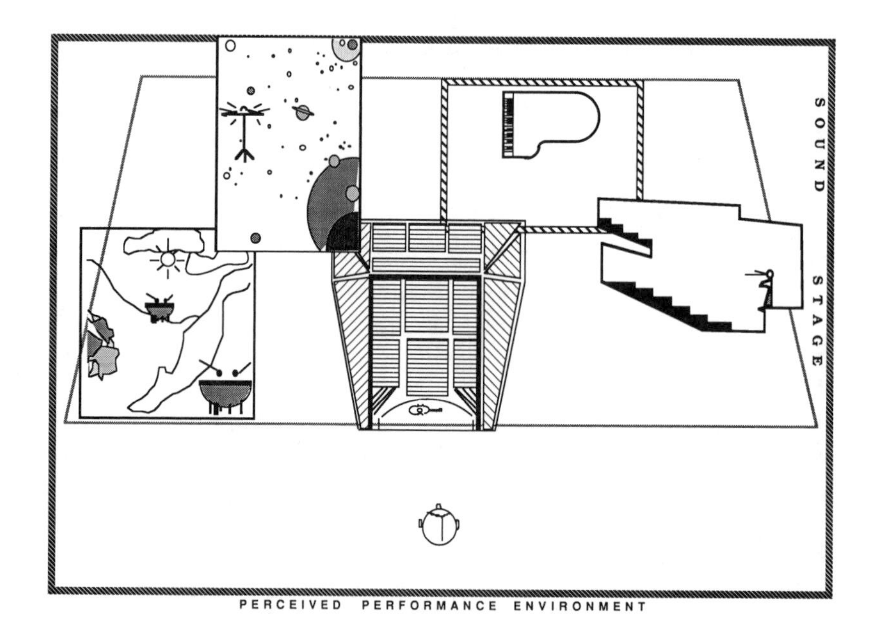

FIGURE 12-3. Space Within Space.

- Timpani located in an open air environment;
- A stringed instrument placed in a large concert hall;
- A vocalist performing in a small performance hall;
- A piano sounding in a small room; and
- A cymbal appearing to exist in a very unnatural (perhaps otherworldly or outerspace), remarkably large environment.

The many simulated acoustical and unnatural environments are perceived as existing within the overall space of the perceived performance environment. Further, more than one source may be placed within an environment, with the distances of the sources contained within the same environment being considerably different.

The simulated acoustical environments of the sound sources and the overall program may be of any size relationship to one another. The host environment of the individual sound source may have the characteristics of a physically large space, and the perceived performance environment may have the characteristics of a much smaller physical environment–this is a common relationship. The spaces of the individual sound sources are understood (by the listener) to be existing within the all-encompassing perceived performance environment, no matter the perceived physical dimensions of the spaces involved.

The spaces of the individual sound sources are subordinate spaces that exist within the overall space of the recording. A further possibility (not commonly used, at present) exists that allows subordinate spaces to appear within other subordinate spaces, within the perceived performance environment. Space within space is a hierarchy of environments existing within other environments; its creative applications have not been fully exploited.

The characteristics of the perceived performance environment function as a reference for determining the characteristics of the individual environments of the individual sound sources. All the environments of a recording will have common characteristics that are created by the perceived environmental characteristics of the perceived performance environment (as discussed above). These characteristics provide a reference for determining the unique characteristics of the individual performance environments of the individual sound sources. The characteristics of the perceived performance environment also function as a reference for determining the distance locations of the individual sound sources within the sound stage.

Distance is perceived as a definition of timbral detail, in relation to the characteristics of the environment in which the sound is produced, and the perceived location of the sound source and the listener within that environment. The listener will perceive the distance of the sound source as it is sounding within its unique environment. The listener will then transfer that distance to the perceived performance environment, adding any perceived distance of the source's environment from the listener's location in the perceived performance environment.

The distance of the sound source from the perceived location of the listener within the host environment of the individual sound source is combined with the perceived distance of that environment from the listener's location in the perceived performance environment, to determine the actual distance location placement of the sound source within the sound stage. Through this process, sound sources (with and within their environments) are conceived at specific distances from the listener.

Placing sounds at a distance and at an angle from the listener (imaging) takes place at the perspective level of the perceived performance environment. Sound sources (with their individual environments and conceived distance locations) will be located at an angle from the listener. The stereo location of the sound sources will place them on the sound stage, within the stereo loudspeaker array, at an angle of direction from the listener. The size of the sound source will be either a narrow and precisely defined point of location or it will be an area between two boundaries. Sources that occupy an area may be of any reproducible width and may be located at any reproducible location within the stereo array. Further, under unusual production practices, it is possible for sound sources to appear to occupy two separate locations or areas within the stereo array. Thus, sound sources (existing in an environment) are presented in two dimensions: at an angle from the listener and at a distance from the listener.

The spatial elements of sound that are universally used in music productions are:

1. The perceived location of the sound source within the stereo array (the left-to-right, horizontal plane, immediately in front of the listener, extending slightly beyond the loudspeaker array);
2. The illusion of a distance of the sound source from the listener, within the perceived performance environment; and
3. The perceived environmental characteristics of the individual environments, in which the individual sound sources appear to exist, in the final recording, and the environmental characteristics of the overall program.

An accurate evaluation of the spatial elements is impossible, unless the listener, listening environment, and playback system function and interact properly. The listener must be located correctly, with respect to the two loudspeakers of the array; the sound system must interact correctly with the listening environment, to complement the reproduced sound and not to distort the signal; and the sound system itself must be capable of reproducing frequency, amplitude, and spatial cues, without distortion.

The perceived elevation of a sound source (location on the vertical plane) and the horizontal localization of sound sources to the sides and behind the listener are not consistently available in widely used audio recording and playback systems. Therefore, they have yet to become a universal resource for artistic expression. Technology and artistic practice will surely continue to develop. These two potential artistic elements related to space may be important components of musical contexts and other audio applications, in the near future.

Additional processes for evaluating spatial characteristics of sound sources, other than those presented in this chapter, will need to be devised, to address the areas of sound source elevation and the horizontal localization of sound sources 360 degrees around the listener, when artistic practice and technology allow these two elements to be found in music productions.

SOUND LOCATION WITHIN THE STEREO ARRAY

Sound location is evaluated within the stereo array, to determine the location and size of the images of the sound sources. These cues will hold significant information, for understanding the presentation of the musical materials and sound sources (the mix of the piece), and may contribute significantly to shaping the musical ideas themselves. Phantom images may change locations or size during a piece of music. These changes may be sudden or gradual and prominent or subtle.

The *stereo sound-location graph* will plot the locations of all sound sources against the time line of the work. The graph portrays the direction of sources from the listener (positioned in the center of the listening environment), as well as the size of the phantom images.

The boundaries of the vertical axis are marked by the "left" and "right" loudspeaker locations (Fig. 12-4). The actual, perceived source locations may extend slightly beyond the loudspeaker locations (up to 15 degrees). Angle is represented by placing the sound source's location within the L-R speaker-location boundaries, assuming a central listening position. Precise degree increments of angle are not incorporated into the graph. The absence of reference distances between loudspeakers and between the loudspeakers and the listener, as well as the absence of a reference angle of the loudspeakers with respect to the listener, make it impossible to establish a universally transferable scale for location placement.

A sound source may occupy a specific point in the horizontal plane of the

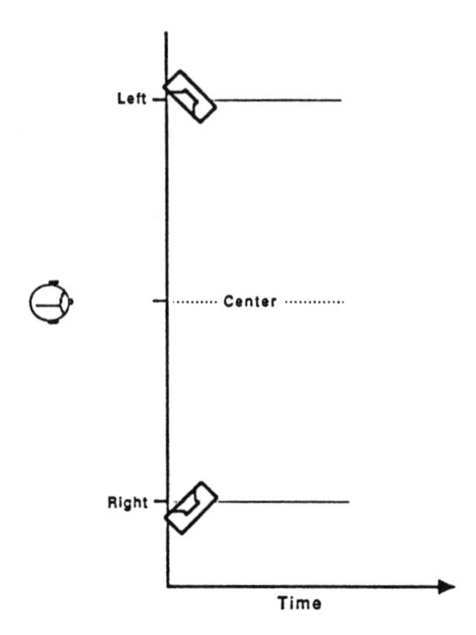

FIGURE 12-4. Continuum for Stereo Sound Location.

sound stage, or it may occupy an area within the sound stage. The graph will dedicate a source line to each sound source, and will plot the stereo locations of each source against a common time line. The source line for point source images will be a clearly defined line at the location of the sound source.

The source line for the spread image will occupy an area of the sound stage, and will extend between the boundaries of the image itself. It is common for stereo sound-location graphs to be multitiered, placing spread images on separate tiers from point sources, or providing a number of separate tiers for spread images. Figure 12-5 presents a single-tier stereo sound-location graph, containing two sound sources: one spread image and one point source.

The process of determining stereo sound location will follow the sequence:

1. During the initial hearing(s), listen to the example, to establish the length of the time line. At the same time, notice the presence of prominent instrumentation, with placements and activity of their stereo location, against the time line.
2. Check the time line for accuracy, and make any alterations. Establish a complete list of sound sources (instruments and voices), and sketch the presence of the sound sources against the completed time line.
3. Notice the locations and size of the sound sources (instruments and voices) for boundaries of size, location, and any speed of changing

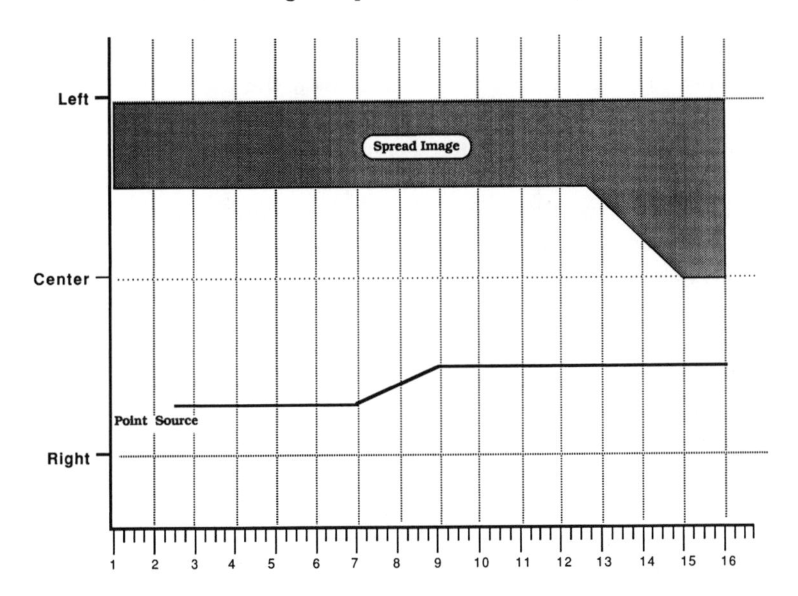

FIGURE 12-5. Stereo Sound-Location Graph.

locations or size of image. Placing instruments against the time line, more than the boundary of speed of changing location, will most often establish the smallest time unit, in the graph, required to accurately exhibit the smallest significant change of location. The boundaries of the sound sources' locations will establish the smallest increment of the "Y" axis required. The perspective of the graph will always be of either the individual sound source or of the overall sound stage.

4. Begin plotting the stereo location of each instrument or voice, on the graph. The locations of spread images are placed within boundaries, which may be difficult to locate during initial hearings, but they can be defined with precision. The listener should continue to focus on the source until it is defined. The locations of point source images are plotted as single lines; these sources are easiest to precisely locate, and are the sources most likely to change locations in real time.

Locations and image size do not often change within sections of a work. Changes are most likely to occur between sections of a piece or at repetitions of ideas or sections (where changes in the mix often occur). The listener should, however, never assume that changes will not occur. Gradual changes in source locations and size are present in many pieces, often in pieces one might not expect to find a sophisticated approach to recording production.

These changes will not be recognized, unless the listener is willing to focus on this artistic element and is prepared to hear these changes.

Continually compare the locations and sizes of the sound sources to one another. This will aid in defining the source locations, and will keep the listener focused on the spatial relationships of the various sound sources. The evaluation is complete when the smallest significant detail has been incorporated into the graph.

The stereo sound-location graph incorporates the following:

1. The left and right loudspeaker locations, a designation for the center of the stereo array, and space slightly beyond the two loudspeaker locations as the "Y" axis.
2. The graph will become unclear if too many sound sources (especially many spread images) are placed on the same tier. The "Y" axis may be broken up into any number of similar tiers (each the same as listed in no. 1), to clearly present the material on a single graph.
3. The "X" axis of the graph is dedicated to a time line that is devised to follow an appropriate increment of the metric grid, or is representative of a major section of the piece (or the entire piece), for sound sources that do not change locations.
4. A single line is plotted against the two axes for each sound source; the line will occupy a large, colored/shaded area, in the case of the spread image.
5. A key will be required to clearly relate the sound sources to the graph. The key should be consistent with keys used for other similar analyses (such as musical balance and performance intensity), to allow different analyses to be easily compared.

Figure 12-6 displays the stereo sound location of a number of the sound sources of The Beatles' work, "A Day In The Life." The location, size, and movements of the sound source images directly contribute to the character and expression of the related musical materials. As an exercise, listen to the recording and notice the placement of the percussion sounds. The stereo locations of the percussion sounds complement the placements of the voice, bass, piano, guitar, and maracas, to balance the sound stage.

DISTANCE LOCALIZATION

Distance localization and *stereo localization* combine to provide the imaging of the sound source. Figure 12-7 is an empty sound stage, onto which sound sources are imagined to be located. Placing sounds on this empty sound stage will allow the listener to make quick, initial observations regarding imaging.

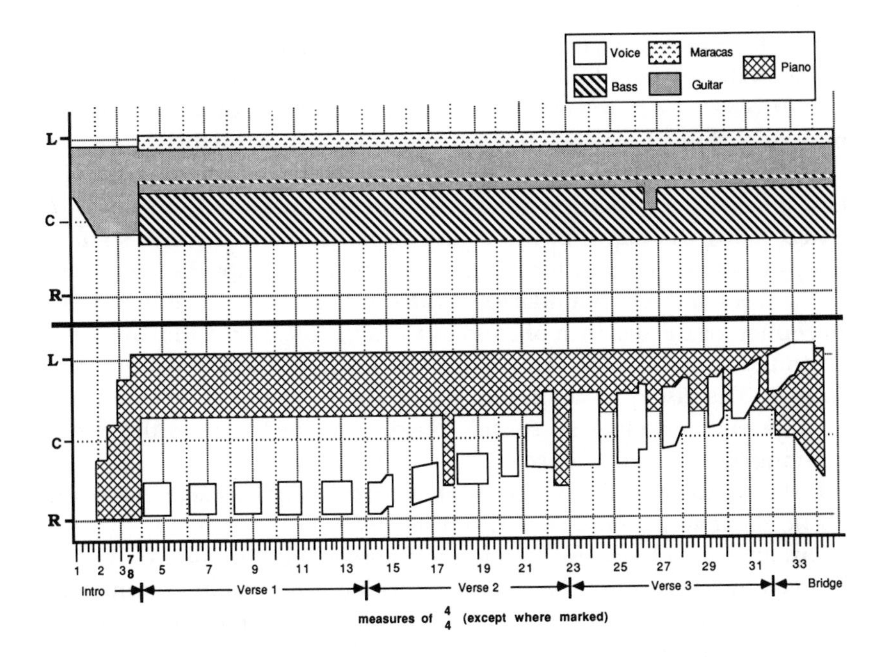

FIGURE 12-6. Multitiered Stereo Sound-Location Graph—The Beatles: "A Day In The Life."

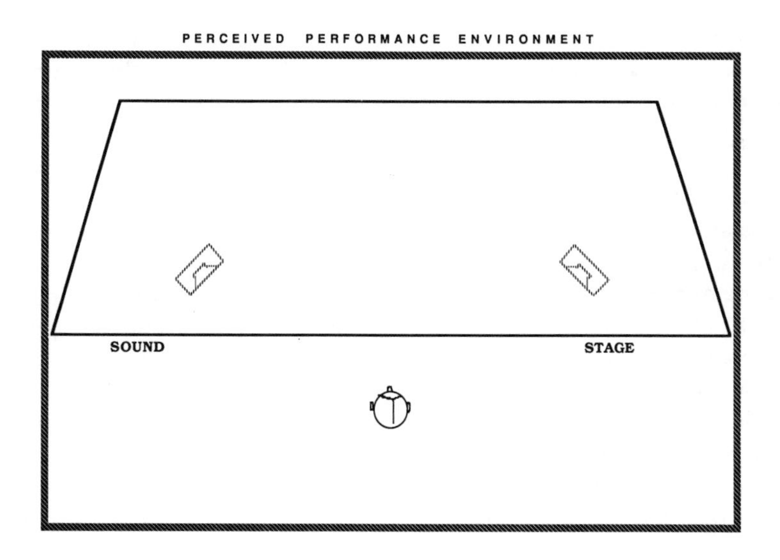

FIGURE 12-7. Empty Sound Stage.

These observations can then lead to the more detailed evaluations of stereo location (above) and of distance (below). It is important to note that these location observations relate to specific moments in time or sections of a work (periods of time). Location changes cannot be written on this figure.

Distance is the perceived location placement of the sound source from the listener. It is a location where the listener imagines the sound to be placed, along the depth of the sound stage. Humans perceive sounds to occupy a precise distance location; sounds do not occupy distance "areas."

The front and rear boundaries of the sound stage are established by the conception of the distance of sources from the listener. The front edge of the sound stage may be immediately in front of the listener, or at any distance. The distance dimensions of the sound stage will be conceived as the area encompassing all sound sources, mapped against the distance location continuum of Figure 12-8.

Distance cues are often not accurately perceived. Many activities of other artistic elements are confused with distance.

Distance is *not* loudness. In nature, distant sounds are often softer than near sounds. This is not the case in recording production. Loudness does not directly contribute to distance localization in audio recordings. At times, loudness and distance cues are associated, but this is not always the case—especially in recording production. A "fade out" may cause sound sources to be perceived as increasing in distance. This perception will be caused by a diminishing level of timbral detail, not by the decreasing dynamic level.

Very often, people will describe a sound as being "out in front," implying a closer distance. The sound may actually be louder than other sounds, or the sound may stand out of the musical texture because of its sound quality. There is much potential for confusing distance with dynamic levels and other aspects of the sound or the musical context that might draw the listener's attention.

Distance is *not* the amount of reverberation placed on a sound source. In nature, distant sounds are often accompanied by a great amount of reverberant energy. Reverberant energy does play a role in distance localization, but not so prominent a role that it can be used as a primary reference. Reverberant energy is most important as an attribute of environmental characteristics, and in placing a sound source at a distance, within the individual source environment. Humans perceive distance within environments, through time and amplitude information extracted from processing the many reflections of the direct sound. The ratio of direct to reflected sound influences distance location. Thus, reverberation contributes to the listener localizing distance, but it is *not* the primary determinant of distance location, in and of itself.

Distance is *not* the perceived distance of the microphone to the sound source that was present during the recording process. The only exception to this statement occurs when the initial recording is performed with a single stereo-pair of microphones, and no signal processing is performed on the overall program. Microphone to sound source distance is a contributor to the timbral characteristics of the sound source. Microphone to sound source distance will determine the amount of definition of the sound source's timbre (how much timbral detail is present in the sound) that was captured by the recording process, and will determine the amount of the sound of the initial recording environment that has become part of the sound source's timbre. Generally, the closer the microphone to the sound source, the greater the definition of timbral components captured during the recording process. This will provide distance localization information, in such a way that very close microphone placement will cause the image to be perceived to be very close to the listener (if no timbral modifications or signal processing is performed in the mix and recording process). The sound quality may be greatly changed in the mix, significantly altering microphone to sound source cues. Microphone to sound source distance contributes to the overall sound quality of the source's timbre. It contributes to the listener localizing distance, through definition of timbral detail, but it is *not* a primary determinant of distance localization, in and of itself.

Distance location is primarily the result of timbral information and detail. Timbre differences between the sound source as it is known in an unaltered state and the sound as it exists in the host environment of the recording are the primary determinants of distance localization. The listener is acutely aware of how timbres are altered over various distances. It is through perceiving these changes that listeners conceive the distance of a source from their listening location. Humans are unable to estimate the physical distance (meters, feet, yards) of a sound source from their location; they perceive distances in relative terms, and compare locations to one another (Fig. 12-8).

Humans rely on timbral definition, for most of their distance judgements, and the ratio of direct-to-reverberant sound to a lesser degree. The extent to which they rely on either factor depends on the musical context. Distance localization is a complex process, relying on many variables that are inconsistent between environments.

The listener knows the sound qualities of sound sources, within their immediate area. The listener perceives him or herself as occupying an area, encompassing a space immediately around him or herself. This area serves as a reference from which to judge "near" and "far." Within this area, sounds have no changes in timbre. All characteristics of timbral content are present, and the sounds will have more detailed definition, the closer they are to the listener. The overall sound quality of the sound source may be somewhat

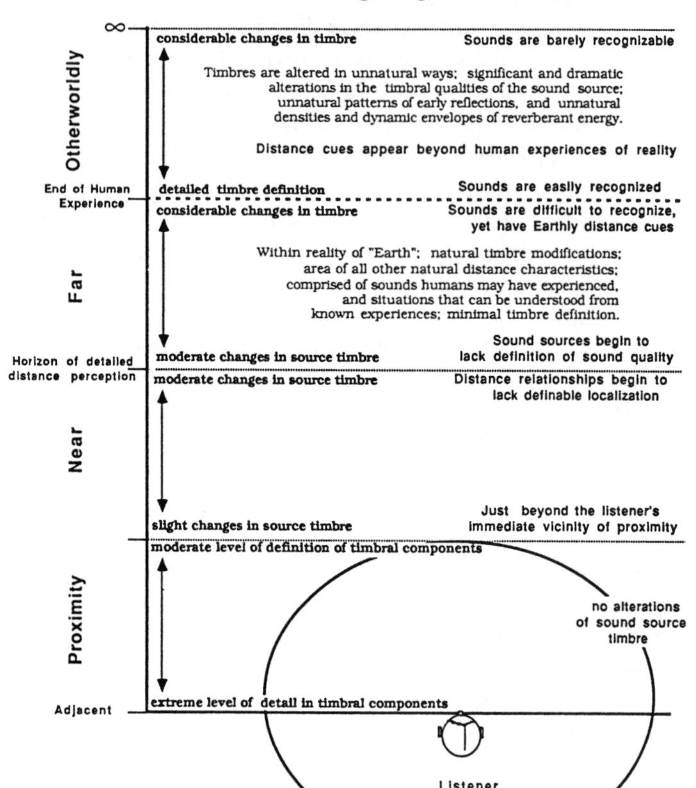

FIGURE 12-8. Continuum for Designating Distance Location.

altered by the characteristics of the host environment, but the level of detail present in the timbre causes the listener to perceive the source as being within the immediate area of *proximity*. This space that immediately surrounds the listener (area of proximity) may be perceived as being rather large or very small, depending on the context of the material and the perspective of the listener.

The listener knows the sound qualities of sound sources at *near* distances. Humans conceive "near" as being immediately outside of their area of proximity. Throughout this "near" area, the listener is able to localize the sound's distance with detail and accuracy. Timbres are very slightly altered in the closest of sounds considered "near," and are moderately altered in the furthest of sounds considered "near." An area will exist between these two boundaries. Sounds cease to be considered "near" when the listener begins to have difficulty localizing distances in detail.

Far sound sources lack sound quality definition. The closest of "far" sounds will have moderate alterations to sound quality, with little or no definition. Few low amplitude partials will be present, and amplitude and frequency attack transients will be difficult to detect. The furthest of "far" sounds will have considerable alterations to sound quality, and the sounds will lack all definition. The furthest "far" sounds may even be difficult to recognize, but they will have natural, or "Earthly," timbral alterations that could occur in nature. An area will exist between these two boundaries of "far"; this area could conceivably be quite large.

Sounds may appear to be located at a distance outside of human experience—a distance beyond our world. These sounds will contain significant alterations to the original timbre, but will remain recognizable. The alterations to the timbre will be such that they could not have occurred in nature. These differences from an unaltered timbre may be frequencies emphasized or de-emphasized that would not normally occur, unnatural changes in the dynamic envelope of the source, or any other unnatural modifications to the sound source. A separate conceptual continuum exists in this *otherworldly* area. Sounds may have the detail of "proximity" (and be located near the lowest boundary of the area) or have very little definition of timbral detail (and be located near infinity).

The listener will initially focus on distance cues of the sound source, at the perspective of the source within its own host environment. The listener will intuitively transfer that information to the perspective of the sound stage. There, the sound source's degree of timbre definition, the perceived distance of the sound source within its host environment, and the perceived distance of the source's host environment from the perceived location of the listener are combined into a single perception of distance location. This process will determine the actual perceived distance of the sound source, at the perspective of imaging. Again, the definition of, or the amount of detail present within the timbre of the sound source, will play the central role in determining perceived distance.

A sound source may lack definition within its host environment, causing the sound source to be localized at a "far" distance within its host environment. The host environment itself may be perceived as beginning at the front of the sound stage (and the front of the sound stage may be adjacent to the listener), but the source's actual location is placed in the "far" area from the listener. Most often, environments are conceived as extending to the front of the sound stage. Distance cues of imaging are then calculated solely from the listener's perceived location in the perceived performance environment that contains the sound stage.

A host environment may be placed at a distance, in the sound stage. The host environment and the sound source(s) within the environment will then show the same alterations of timbral components; these alterations are caused by the perceived performance environment. The relationship of the distance of the source within its environment will be unchanged, but the distance of the source to the listener's location will be perceived as altered. Sounds may thus be close, within their own environment, yet be given distance cues, to move the sound (within its environment) in distance. This concept is only effective at increasing distance.

It is possible for sounds to have detailed definition of sound quality, and appear close to the listener, yet be in an environment that is otherworldly. Information that could not occur in reality may be present in the environmental characteristics of the sound source, no matter what the definition level of the sound source timbre. Environmental characteristics alone may place the sound in the otherworldly area. The sound may be localized at a close distance to the listener (perhaps adjacent to the listener), and placed at or near the lowest otherworldly boundary. The listener would perceive the sound to be occurring within (perhaps imagine themselves to be extended into) an environment that is physically impossible within human reality. The listener's perceived location would simultaneously be at the edge of "proximity" and at the edge of the "otherworldly" area.

The continuum for distance location extends from "adjacent" to "infinity." Adjacent is that point in space that is immediately next to the space that the listener is occupying. It should be conceived literally as being the next molecule available beside the listener, as a sound may be localized at that location.

The continuum for distance localization consists of four areas. The areas represent conceptual distance, not physically measurable distance increments. Distance is judged as a concept of space between the sound source and the listener. An area of "proximity" surrounds the listener; this area serves as a reference for judging "near" and "far" distances. Human experience of the nature of sound is used as a reference, to conceptualize the amount of space (distance) between the source and the listener.

The four areas of the continuum are:

1. An area of "proximity," the space that the listener perceives as his or her own area. This area immediately surrounding the listener may be extended to be conceived as a small to moderately-sized room around the listener. The listener will perceive the "proximity" area as being his or her own immediate space.
2. A "near" area, as the area immediately outside the space that the listener

perceives him or herself as occupying, extending to an horizon where the listener begins to have difficulty localizing distances in detail.
3. A "far" area, beginning where perception dictates space ceases to be "near" (where detailed examination of the sound is difficult) and extending to where sounds are almost impossible to recognize. Extreme "far" sound sources contain very little definition of sound quality.
4. An "otherworldly" area, where sounds appear to be at a distance outside of human experience or are located within environments that cannot exist on Earth. The area extends from "infinity," or a very great distance (perhaps a distance beyond our world), to a secondary "adjacent" location within this "Other World." In this entire area, sounds are recognizable and timbre alterations are unnatural.

Reverberation is often more important than timbral definition in determining the existence of the "otherworldly" area. Distance location placement remains largely defined by timbre detail, although ratio of direct-to-reverberant sound will play more of a role than in the other three areas. The concept of the area of "otherworldly" is based on unnatural environmental characteristics. Timbre definition is the primary determinant of distance location in the other three areas.

These four areas are not of equal size. The amount of physical distance contained in the conceptual area of "proximity" will be considerably different than the physical distance encompassed by the conceptual area of "far." All four areas of the continuum occupy a similar amount of "conceptual" space, but represent significantly different amounts of physical area. The vertical axis of the *distance location graph* must clearly divide the four areas.

The size of the four areas may be adjusted between appearances of the distance location graph. The amount of vertical space occupied by the areas may be adjusted to best suit the material being graphed, with certain areas being widened in certain contexts and narrowed in others. The areas of "otherworldly" or "far" may even be omitted in certain contexts. The area of "proximity" should always be included (although it may be narrowed to occupy less vertical space, if necessary), to clearly present the conceptual distance, between the perceived location of the listener and the front edge of the sound stage.

Sound sources will be placed on the graph:

1. By evaluating the definition of the sound quality of each sound source (the amount of detail present in the timbre of the sound source), and by evaluating the ratio of direct-to-reverberant sound and the quality of the reverberant sound; and
2. By directly comparing the sound source to the perceived distance

locations of the other sound sources present in the musical context (using proportions of different locations between three or more sound source distances, to make more meaningful comparisons).

The individual listener's knowledge of timbre and environmental characteristics and his or her ability to recognize the sound source are the variables that may cause the listener to inaccurately estimate distance. A very close sitar may sound like a "far" sound, to a person who does not know the sound of a sitar. As the sets of life experiences of listeners varies, so does the individual's ability to conceptualize the distance relationships of sounds.

Distance judgements are difficult to conceive and perceive, during initial studies. Distance is, however, a central concern of sound source imaging, and thus of music production. Skill in this area can be refined and should become highly developed.

Distance judgements may have a somewhat subjective element in placing sounds within specific areas. This is especially true for the furthest two areas. The placing sources at a distance within the "otherworldly" area may be somewhat speculative and inconclusive, with different, equally valid placements possible, between individuals.

Determining distance location will follow this sequence:

1. During the initial hearing(s), establish the length of the time line. Notice the primary instrumentation and any prominent placements and activity of distance location, especially as they relate to the time line.
2. Check the time line for accuracy, and make any alterations. Establish a complete list of sound sources (instruments and voices), and sketch the presence of the sound sources against the completed time line.
3. Make initial evaluations of distance locations. Notice the locations of the sound sources, to establish boundaries of the sound stage (the location of the front and rear of the sound stage). Notice any changing distance locations, and calculate any speed of changing locations. The placement of instruments against the time line, more than the boundary of the speed of changing location (which are quite rarely used), will most often establish the smallest time unit required in the graph to accurately exhibit the smallest significant change of location. The amount of activity in each area will establish the amount of "Y" axis-space required to plot the area's sound sources. The perspective of the graph will always be of either the individual sound source or the overall sound stage.
4. Begin plotting the distance of each instrument or voice on the graph. Sound sources will be placed on the graph by the following procedure:
   a. Evaluate the definition of the sound quality of each sound source, by focusing on the amount of detail present in the timbre of the sound

source, being aware of the amount and characteristics of the reverberant sound.

b. Temporarily transfer this evaluation into a distance of the source from the listening location, and pencil-in the sound on the distance location continuum for reference.
c. Reconsider the definition of the timbre: the definition realized in b. will often be the distance of the sound source, within the source's host environment. That distance must then be conceived in relation to the sound stage.
d. Evaluate the distance of the source's host environment to the listening location in the perceived performance environment, and calculate a location of the front edge (often, this distance will not be present).
e. Combine the distances of d. and c. (the distance of the source within its own host environment), to arrive at the distance location of the sound source in the perceived performance environment.
f. Precisely locate the distance location of the sound source by comparing the sound source's location to the locations of other sound sources.

The locations of all sources are plotted as single lines. Sources are precisely located at a specific distance from the listener. Sound sources do not often change locations in real time, or within sections of a work. Changes usually occur between sections of a piece, at entrances or exists of individual sound sources, or at repetitions of ideas or sections (where changes in the mix often occur). The listener should, however, never assume changes will not occur. Gradual changes in distance are present in many pieces.

Once several sounds are accurately placed on the distance continuum, distance location is most readily accomplished by directly comparing the sound source to the perceived distance locations of the other sound sources present in the music. Use proportions of differences between the locations of three or more sound source distances, to make for more meaningful comparisons. Is sound "c" twice or one-half the distance from sound "a" as "a" is from sound "b"? How does this compare to the relationship of sounds "d" and "c"? Sounds "c" and "b"? Continually compare the distance locations of the sound sources to one another. The evaluation is complete when the smallest significant detail has been incorporated into the graph.

The *distance location graph* incorporates the following:

1. A continuum from "adjacent" through "infinity" (divided into four areas) as the "Y" axis.

2. The "X" axis of the graph is dedicated to a time line that is devised to follow an appropriate increment of the metric grid.
3. A single line is plotted against the two axes for each sound source.
4. A key will be required, to clearly relate the sound sources to the graph. The key should be consistent with keys used for other similar analyses (such as musical balance or stereo location), to allow different analyses to be easily compared.

Figure 12-9 is a distance location graph from The Beatles' "A Day In The Life." The voice part has been placed in the "otherworldly" area because of its unnatural reverberation qualities. Its timbral detail might otherwise place the source in the rear-third of the "proximity" area. The sound sources have greatly varied distance locations, giving the sound stage great depth. The guitar, maracas, and bass are in the "proximity" area, and the piano is at the front of the "near" area. As an exercise, place the percussion sounds in an appropriate distance location. Notice the great conceptual distances between the various instruments of the drum set, as some sounds are located at the rear of the "far" area.

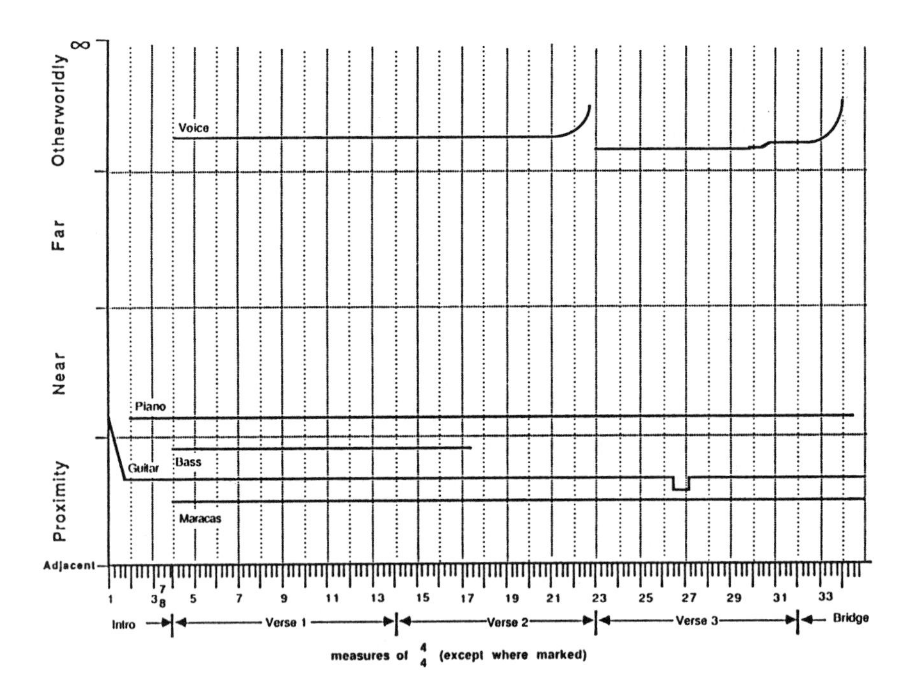

FIGURE 12-9. Distance Location Graph—The Beatles: "A Day In The Life."

ENVIRONMENTAL CHARACTERISTICS

The characteristics of the sound source's host environment are important in shaping many qualities of the recording. They contribute significantly to the overall quality of the sound source. They likewise have a significant influence on the ways that the sound source and its musical materials are perceived within the musical context of the work. They largely define the illusions of the content of the perceived performance environment, of space within space, and of the imaging of the recorded piece of music.

The environmental characteristics of the entire program (the perceived performance environment) shape the conceived space in which a performance is occurring. The characteristics of the conceived space of the perceived performance environment will greatly influence the conceptual setting for the artistic message of the work.

Environmental characteristics of both the host environments of the individual sound sources and the perceived performance environment play significant roles in music production. These artistic elements have the potential to provide significant information for communicating the musical message of the piece of music. Currently, they are most often used in supportive roles, coupled with sound quality in defining the unique characters of individual sounds, and as a separate element in creating depth of sound stage, in creating a varied set of resources for space within space, and in creating the illusion of the perceived performance environment.

The recordist is concerned with recognizing the characteristics of the environments within which the sound sources exist, the characteristics of the perceived performance environment, and the influence of the environment on the overall sound qualities and on the effectiveness of the sound sources and its musical materials.

The listener is presented with a composite sound of environmental characteristics. The characteristics occur by sounding a sound source within the environment. The sound source and the environment interact and create a composite sound, a new overall sound quality. To understand the influence of the host environment on the sound source, in making this composite sound, the environment and the sound source must be evaluated separately.

Listeners perceive environmental characteristics as an overall sound quality that is comprised of a number of component parts. As global sound qualities, environmental characteristics are conceptually similar to timbre. The evaluation of timbre (sound quality) and environmental characteristics will be similar, in that both will seek to describe the states and activities of the physical components of sound. While we recognize large halls, small halls and other spaces as having common environmental characteristics, each environment is unique. For people to adequately communicate about environmental characteristics, the environment must be defined by its unique sound characteristics. These characteristics can be objectively described only by discussing the states and activities of the components parts of the environment's sound.

Environmental characteristics appear as alterations to the sound source's timbre, created by the sound source and the environment interacting. The evaluation of the environment's character is thus a definition of the changes that have occurred in the sound source's timbre, after being placed within the environment.

Environmental characteristics are determined by the listener, through comparing his or her memory of the sound source's timbre outside of the host environment to the sound source's timbre within the host environment. The listener must go through this comparison process carefully, scanning the composite sound for information and then comparing that information with his or her previous experiences with the timbre of the sound source, as well as with other environments. Differences in the spectrum and spectral envelope of the sound source, as remembered by the listener, and as heard in the host environment, form the basis for determining most environmental characteristics.

If the listener does not recognize the timbre, or has no prior knowledge of the sound source, he or she will be at a disadvantage when calculating the characteristics of the host environment. There will be no point of reference for determining how the environment has altered the timbre of the original sound source. The listener must rely on knowledge of similar sounds and on knowledge of the characteristics of similar environments and must calculate estimations of the characteristics of the environment.

The listener is seeking to define the characteristics of the environment itself. The listener must make certain that he or she is *not* identifying characteristics of the sound source, and must make certain that he or she is *not* identifying characteristics of the sound source within the environment. The characteristics of the environment have specific component parts that contribute to its own unique sound quality. These characteristics are what must be determined, by identifying the differences between the sound quality of the sound source itself and the sound quality of the sound source within the environment. The component parts of environmental characteristics are: (1) the "reflection envelope," (2) the "spectrum," and (3) the "spectral envelope." The *environmental characteristics graph* (Fig. 12-10) allows for the detailed evaluation of these three components.

The "reflection envelope" is created by the amplitudes of the initial reflections and the reverberant energy of the environment, throughout the duration of the environment. This envelope is comprised of many reitera-

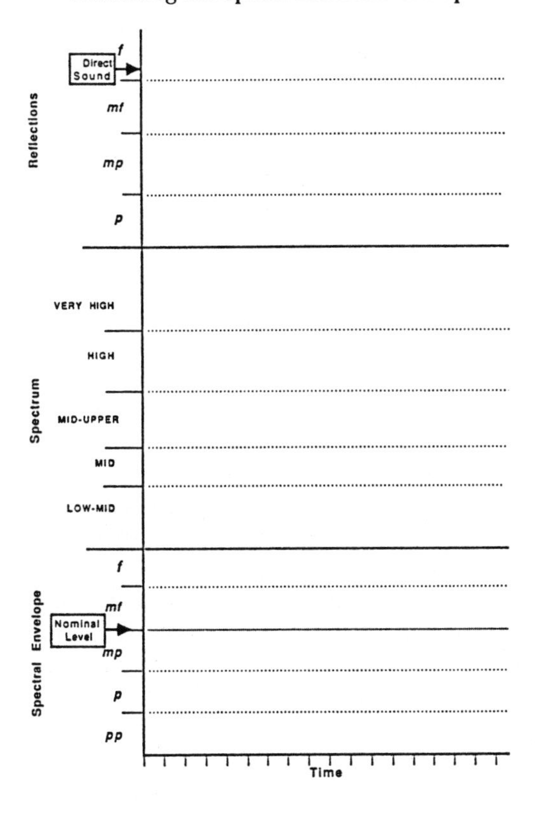

FIGURE 12-10. Environmental Characteristics Graph.

tions of the sound source. The reiterations vary in dynamic level and in spacing between one another (time density).

The spacing of the reiterations of the sound source will be dramatically different over the duration of the sound of the environment (for example, the spacing of the early reflections will be considerably different from the spacings of reflections near the end of the reverberant energy). This portion of the graph must clearly show the time of arrival of the reflected sounds to the listener location, the density of the arrival times of the reflected sounds, and the amplitude of those arrivals, in relation to the amplitude of the direct sound.

The amplitude of the direct sound is designated as a reference level for calculating the dynamic levels of the reflected sound. The recordist should use his or her skills at pattern recognition, to extract time information from the sound of the environment. The reflections portion of the graph will show the following information. The listener should organize his or her listening to the time elements of the environmental characteristics to recognize this information:

- Patterns of reflections created by dynamics;
- Patterns of reflections created by spacings in time;
- Spacing of reflections in the early time field;
- Dynamic contour of the entire reflections portion;
- Density (number and spacings of reflections) of reverberant sound;
- Dynamic relationships between the direct sound, individual reflections (of the early time field), and the reverberant sound; and
- Dynamic contour shapes within the reverberant sound.

An isolated appearance of the sound source in the host environment must be found for all the time and reflection-amplitude information to be accurately evaluated. This is especially true for the decay of the reverberant energy and the spacings of the early reflections (which play key roles in the characterization of the environment). A short (staccato) sound, from the sound source, will allow the reflection information to be most audible, since it will not have to compete with the sound of the source itself.

Without the opportunity to hear the environment's complete presentation, information related to the reverberant energy might never be audible. The listener should try to find several appearances of each sound source were they can be heard alone, without other sound sources, and where they are playing short durations.

Many hearings of the sound in a wide variety of presentations will be necessary to compile an accurate evaluation of all the time and amplitude characteristics of the environment. Further, some pitch/frequency information can be obtained only by listening to the environment, as the sound source performs many different pitches. This allows the emphasized and de-emphasized frequency areas of the environment to be heard.

The "spectrum" of the reverberant sound and the initial reflections is a composite of the frequencies or bandwidths of pitch areas that are emphasized and de-emphasized by the characteristics of the environment itself. This spectrum will be only those frequencies that are affected by the environment. The environment may emphasize or de-emphasize bandwidths of frequencies, or specific frequencies. Often, the spectrum of the environment will contain only a small number (three to seven) of prominent frequencies or pitch areas that are either emphasized or de-emphasized by the environment itself.

These frequencies are determined by carefully evaluating many appearances of the sound source in the environment and by listening to the way the sound source's timbre is changed by the environment over a wide range of pitch levels. Some appearances of the sound source will not have frequency information in certain frequency areas that may be emphasized or de-emphasized by the environment. The listener must scan many pitch levels of the sound source, to determine the spectral content and the spectral envelope of the environment.

The "spectral envelope" of the environment is how the frequencies that are emphasized and de-emphasized by the characteristics of the environment itself (spectrum) vary in loudness level, over the duration of the sound of the environment. The "spectral envelope" and "spectrum" portions of the graph are coordinated in showing different activity of the same sound components (as with sound quality evaluation).

A *nominal level* is established and transferred to the environmental characteristics graph as a reference for plotting the dynamic contours of the spectral components. The nominal level will vary in direct relationship with the dynamic envelope of the environment (reflections envelope). The nominal level *is* the dynamic envelope of the environment, where the sound source's frequency components are unaltered. This dynamic envelope is transferred into a conceived steady-state on the "spectral envelope" portion of the graph. The dynamic envelope of the environment changes over time; it is the dynamic contour that is outlined by the "reflections envelope."

The nominal level is placed at the dynamic level, precisely between mezzoforte and mezzopiano. Frequencies or pitch areas that are emphasized by the environment will be plotted as activity above the nominal level. Frequencies or pitch areas that are de-emphasized by the environment will be plotted as activity below the nominal level.

Determining environmental characteristics will follow this sequence:

1. During the initial hearing of the entire work, listen to the sound source, to identify a location where the sound is isolated throughout the duration of the environment. The time increments of the time line will often be milliseconds. The calculation of such small time increments will take considerable practice (working with a delay unit altering a short percussive sound will assist in developing time estimation skills of these small time units). Establish the length of the time line of the environmental characteristics by timing reflections of the early time field and the reverberant sound. This is most easily accomplished for short percussive sounds, but environmental characteristics evaluation is possible for any sound source, if the reverberant energy of the environment is exposed (not accompanied by other sound sources), after the sound source has stopped sounding.
2. Check the time line for accuracy, and make any alterations. Work in a detailed manner, to establish a complete evaluation of the reflections

of the sound. First, sketch the presence of the most prominent reflections against the completed time line; then, establish the precise time placement and the dynamic levels of these prominent reflections against the time line. Use the prominent reflections as references, to fill in the remaining reflections in the early time field. After the early time field is plotted, complete the reflections portion of the graph by plotting the dynamic envelope and spacing of reflections (density) of the reverberant energy.

3. Notice the locations and size of any emphasized or de-emphasized pitch areas or frequencies. Scan the entire piece of music, listening to the sound source and the environmental characteristics through many different pitch registers. Throughout these listenings, keep track of pitch areas or specific frequencies that appear to be emphasized or de-emphasized. With a running list of observations, regularly identified pitch areas/frequencies will begin to emerge. Further hearings will allow the listener to more accurately identify these frequencies and pitch areas (that make up the spectrum of the environmental characteristics) and to place the presence of these frequencies or pitch areas against the time line.
4. The listener will now plot the dynamic contours of the components of the spectrum against the time line. This process is the same as the process of plotting the spectral envelope of sound quality evaluations. Each component of the spectrum is plotted as a single line. These components are listed in a key so that their dynamic contours may be related to the spectral envelope tier of the graph.

Continually compare the dynamic levels and contours of the spectral components to one another. This will aid in defining the nominal dynamic level (where the amplitudes of the spectral components of the sound source are unaltered by the environment), will aid in keeping the dynamic levels and contours consistent between spectral components, and will keep the listener focused on the relationships of the sound source and its host environment. The evaluation is complete when the smallest significant detail has been incorporated into each tier of the graph.

The *environmental characteristics graph* incorporates the following:

1. Three tiers as the “Y” axis: “reflections” (a continuum of dynamic level), “spectrum” (a continuum of pitch level), and “spectral envelope” (a continuum of dynamic level).
2. The “reflections” portion of the graph is comprised of a vertical line at each point in time when a reflection occurs; the height of the vertical

line corresponds to the amplitude of the reflection. The dynamic level of the direct sound is indicated on the vertical axis and serves as a reference for calculating the dynamic levels of the reflections. This portion of the graph presents information on the dynamic contour of the reflections of the environment, as well as the spacings, in time, of the reflections throughout the sound of the environment.

3. The "spectrum" portion of the graph is comprised of the registers established in Chapter 9. Spectral components are placed against the "Y" axis by pitch/frequency level. A single line is plotted against the two axes for each spectral component: the line will occupy a large, colored/shaded area, in the case of pitch area, and a narrow line, in the case of a specific frequency/pitch level.
4. The "spectral envelope" portion of the graph depicts the dynamic contours of the spectral components, using dynamic areas as the "Y" axis.
5. The "X" axis of the graph is dedicated to a time line that is devised to incorporate an appropriate time increment, to clearly display the smallest change of a duration, dynamics, or pitch present in the characteristics of the environment.
6. A key will be required to clearly relate the components of the "spectrum" and "spectral envelope" tiers of the graph.

The perspective of the environmental characteristics graph will always be of either the individual sound source, or of the perceived performance environment.

It is not always possible to compile a detailed evaluation of environmental characteristics. The information of the environment is often concealed by the other sounds in the musical texture and is not easily separated from the sound quality of the sound source itself. The ability to recognize environmental characteristics involves much practice. It relies on a knowledge of many sound sources, an ability to evaluate sound quality of the sound source within the host environment, and an acquired skill for comparing and contrasting a previous knowledge of the sound source, with the appearance of the sound source within the environment that is to be defined.

The complexity of environmental characteristics may vary widely. Certain environments will have very few frequency differences from the original sound source. Some environments will not have different time information between the early time field and the reverberant energy, and will have the reverberant energy increase in density through a simple, additive process. Other environments may be quite sophisticated in the way they were created: with time increments of the early time field precisely calculated at different time intervals, with spectral components precisely tuned in patterns of frequencies, designed to complement the sound source, and with spectral envelope characteristics reacting accordingly.

It is possible for all perceived environmental characteristics to be changed in real time, with current technology. It is also possible for the perceived environment of a sound source to be generated solely by a delay unit, by a simple reverberation unit, or by any similar process that would provide easily calculated cues. Although these environments would not be perceived as natural spaces, the listener would proceed to imagine an environment created by the impressions of those simple characteristics.

Figures 12-12 and 12-13 (of the next section) present several environmental characteristics evaluations.

SPACE WITHIN SPACE

The overall environment of the program provides a setting within which the subordinate environments of the individual sound sources will appear to exist. This overall environment is a constant that equally influences the individual environments of all sound sources.

The overall environment is either (1) perceived by the listener as being a composite of the dominant, predominant, and/or common characteristics of the individual environments of the sound sources, or (2) is a set of environmental characteristics that is superimposed on the entire program.

Works will be perceived to have a single overall environment that is present throughout the piece. When this overall environment is created by adding environmental characteristics to the entire musical texture, it is possible for the overall environment to change during the course of a work. In such instances, abrupt changes (usually at a major division of the form of a piece, such as between verse and chorus) are most common. The various environments will be perceived as having occurred within a single overall environment, even though a single environment is not present.

This overall environment, or the perceived performance environment, may be a composite of many perceived environments and environmental cues. The listener will perceive the overall performance environment in this way, if an overall environment has not been applied. In these instances, the perceived performance environment is conceived by the listener through environmental characteristics that are (1) common or complementary, between the environments of the individual sound sources, (2) prominent characteristics of the environments of prominent sound sources (source that presents the most important musical materials, or the loudest, nearest, or furthest sound sources, as examples), and/or (3) the result of environmental characteristics found in both of these areas.

Within this overall environment, the individual environments of the individual sound sources are perceived to exist. This is the illusion of space within space. If reverberation has been applied to the overall program, to create a perceived performance environment, all the sound sources and their host environments, in the recording, will be altered by those sound characteristics.

Distance location and environmental characteristics are related. The interrelationships of distance location and space within space should always be considered when evaluating a music production. The two are closely interactive, and comparing the two will offer the listener many insights into the creative ideas of the recording.

The distance location graph is not the most appropriate place to designate space within space relationships. It often lacks detail of activity and precision. Further, it usually does not present information that is as musically significant as that presented by the musical balance graph. Also, the distance location graph is more difficult to quickly and accurately create (with the exception of advanced individuals) than the musical balance graph.

The musical balance graph is used to record the relationships of the individual environments of the sources. The individual sound-source lines of the musical balance graph are accompanied by numbers ("E1," "E2," etc.). These numbers will correspond with specific environmental characteristics evaluations. The environmental characteristics evaluations are recorded separately, in a master listing.

The listing serves as a key for understanding the characteristics of the environments of each individual sound source included on the musical balance graph. The master listing begins with an evaluation of the environmental characteristics of the overall program. It then lists all the individual environments used in the work. It is possible that more than one instrument is present in the same environment—the sources may be at different distances within the same environment.

Determining space within space follows this sequence:

1. Identify the various environments of the piece. Some sound sources may share environments with other sound sources (at the same or different

distances); some sources may change environments several times in the piece.

2. Perform general environmental characteristics evaluations of the environments. These initial evaluations should be general in nature, seeking prominent characteristics rather than detail.
3. Compare the environments for similarities of time, amplitude, and frequency information. This observation will determine common traits between the individual environments of the sources. These common traits will signal a possible applied, overall environment, if they are present in all environments equally. If the common traits are not applied to all sources equally, they are combined with some predominant traits from the environments of musically significant sound sources, to directly contribute to the characteristics of the overall environment.
4. Listen to the work again, to identify an overall environment of the program. An applied overall environment will be most easily detected by its detail in spectral changes of the reverberant sound, and in the clarity of the initial reflections of the early time field. The characteristics of these environments will be perceived by listening for detail at a close perspective of slight changes to the predominant characteristics of the environment. Overall environments that are an illusion, created by the composite and predominant characteristics of the individual sound sources, will have characteristics that are not readily apparent. The characteristics of these environments will be perceived by listening at the more distant and general perspective of the dominant characteristics of the environment.

   Compile a detailed environmental characteristics evaluation of the overall environment. The evaluation is complete, when the smallest significant detail has been incorporated into each tier of the graph. Begin the master listing of environments with this overall environment.

   If the overall environment changes during the work, the above four steps will need to be repeated at each change of environment. These changes are easily denoted on the time line of the work.
5. Perform detailed environmental characteristics evaluations of the individual host environments of each sound source. The characteristics of the overall environment may or may not be present in these evaluations, depending on the nature of the overall environment and the nature of the individual sound sources' environments. The evaluation is complete when the smallest significant detail has been incorporated into each tier of the graph.

   Number each environment, and enter the evaluation to the master listing of environments. Note on the master listing the sound source or sources that are present within the environment.
6. Create a musical balance graph of the music. Designate the environment

number, next to the appropriate sound source, on the musical balance graph, to create a *musical balance/space within space graph*. Each sound source on the graph should be accompanied by a number, designating a corresponding environmental characteristics evaluation, to be found on the master listing.

Tracy Chapman's "Fast Car" has been plotted in Figures 12-11 through 12-13, demonstrating musical balance/space within space and environmental characteristics. The overall environment is consistent throughout the song; the environments of a number of instruments change between the choruses and the remainder of the piece. As an exercise, identify the other environments of the piece and the other musical balance relationships. Evaluate the environments, being especially careful of the relationships between the dynamic levels of the direct sound versus early reflection and reverberant sounds. Determine whether the overall environment is an "applied" or a "composite" environment.

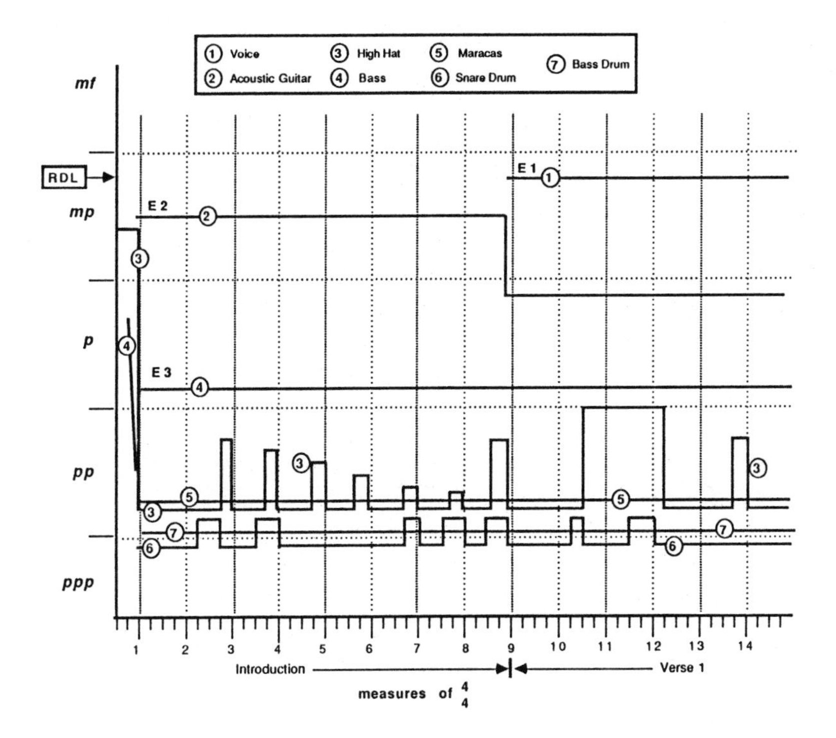

FIGURE 12-11. Musical Balance/Space Within Space Graph—Tracy Chapman: "Fast Car."

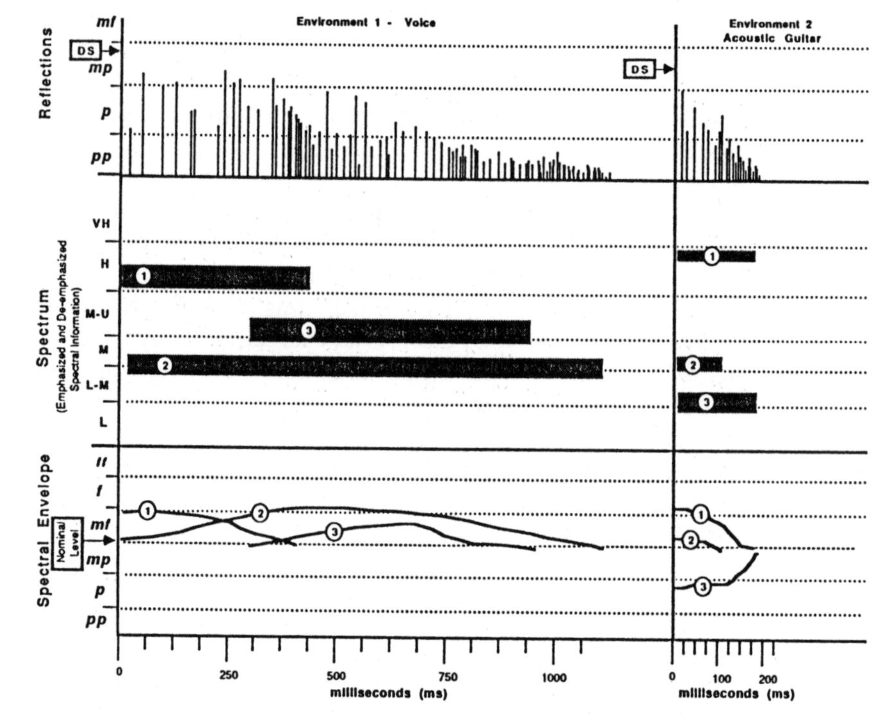

FIGURE 12-12. Environmental Characteristics Graphs—Tracy Chapman: "Fast Car"; Environments #1 and #2.

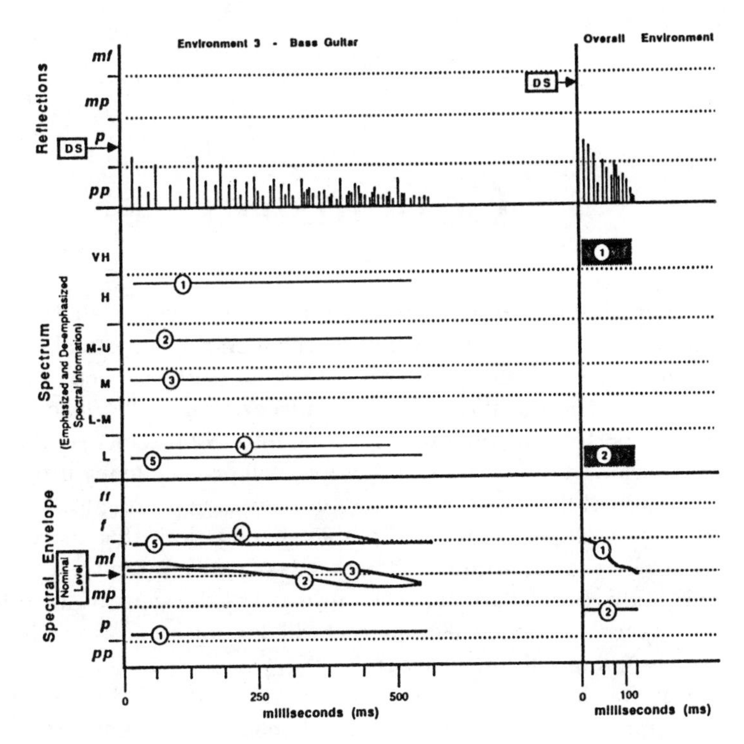

FIGURE 12-13. Environmental Characteristics Graphs—Tracy Chapman: "Fast Car"; Environment #3 and Overall Environment.

Conclusion

The listener should perform the exercise of evaluating all the artistic elements, the traditional musical and text elements, in a particular piece of music. This project will allow the listener to explore, in great depth, the inner workings of all sound relationships of the piece of music. The listener would benefit from performing this exercise on a number of pieces, over the course of a long period of time. These evaluations will provide many insights into the unique use of the musical materials of recording productions.

The project of performing a complete evaluation of a piece will be lengthy. It will take the beginner many hours of concentrated listening. The demands of this project are, however, justified by the value of the information and experience gained. This project will develop and refine critical and analytical listening skills in all areas.

An entire work should be evaluated. The listener should evaluate all the artistic elements, the traditional musical materials, and the text. Upon evaluating these aspects individually, the listener should note how these aspects relate to one another, and how they enhance one another.

This complete analysis of an entire recording is strongly encouraged. Many aspects of a recording will become evident only when evaluations of several artistic elements are compared with one another and to the traditional musical materials. The various roles of the artistic elements in communicating the musical message of the work will become much more apparent, when their interrelationships are recognized. To observe only a small amount of the information that would be determined by such an in-depth study, note the interrelationships of the sound sources in the distance location and stereo location graphs of The Beatles' "A Day in the Life" (Figures 12-6 and 12-9); note how those activities relate to the musical materials the sources are presenting.

The listener will compile a large set of data, while performing the many evaluations of Chapters 9 through 12. These many evaluations will represent many different perspectives and areas of focus. Some of this information will be pertinent to understanding the musical message of the work, some of the information will be pertinent to understanding the qualities of sound in the work (such as how stereo location is used in presenting the various sound sources), and some of the information will be pertinent to appreciating the technical qualities of the recording. All this information will contribute to the listener's complete understanding of the piece of music, and to how the piece made use of the recording medium.

The sequence of evaluations that may prove most efficient in performing an analysis of the entire work (depending upon the individual work—it may vary slightly) is:

- List all the sound sources of the recording.
- Evaluate the pitch areas of unpitched sounds.
- Define unknown sound sources through sound quality evaluations.
- Create a time line of the entire work.
- Plot each sound source's presence against the time line.
- Designate major divisions in the musical structure, against the time line (verse, chorus, etc.).
- Mark recurring phrases or musical materials, similarly, against the time line; an in-depth study of traditional musical materials would be appropriate at this stage.
- Evaluate the text for its own characteristics and its relationships to the structure of the traditional musical materials.
- Perform any necessary melodic contour analyses of those lines that fuse into contours.
- Perform a program dynamic contour evaluation.
- Perform a musical balance evaluation.
- Perform a performance intensity evaluation.
- Evaluate the work for stereo location.
- Evaluate the work for distance location.
- Perform environmental characteristics evaluations of all host environments of sound sources and of the perceived performance environment.
- Notate the space within space information, on the musical balance graph created above.
- Make observations on the interrelationships of all the above.

The following materials may be coupled on the same graph (on separate tiers) or on similar graphs. They are all at the same perspective (at the level of the sound source):

- Performance intensity;
- Musical balance;
- Distance location;
- Stereo location; and
- Space within space.

These five artistic elements will be interrelated, in nearly all musical productions. Observing the interrelationships of this data will allow the listener to extract significant information on the recording. In making these observations, the listener will continually formulate questions about the recording, and seek to find solutions to those problems.

The questions of "how" artistic elements (and all musical materials) relate to one another will center around:

- Patterns of activity within any artistic element (patterns of activity are sequences of levels within the artistic element, and rhythmic patterns created by the relationships of those levels);
- States of value (how high-pitched, what loudness levels, etc.) of any artistic element; and
- Interrelationships of patterns between artistic elements (do the same or similar patterns exist in more than one element?).

Music is constructed as similarities and differences of values and patterns of musical materials. This is also the way that music is perceived by humans. They will perceive patterns within the music (its materials and the artistic elements); they will perceive the qualities (values) of the elements of the music; and they will relate the various aspects of the music to one another.

At the same time, they will compare what they are hearing with what they have previously heard, as well as to their previous experiences. They will attempt to make sense of this information by looking for similarities and differences between the materials.

Listeners will ask, "what is similar" between two musical ideas (or artistic elements), "what is different," and "how are they related?" They will seek answers, by observing the information that was collected through the many evaluations, above. The shapes of the lines on the various graphs may show patterns, and the vertical axes of the graphs may show the extremes of the states of the materials and all their other values.

The listener's ability to formulate meaningful questions for evaluating and analyzing the piece of music (or sounds in general) will be developed over time and with practice. The listener will be asking: "what makes this piece of music unique," "how is this piece of music constructed," and "what makes this recording effective?" Many other, much more detailed questions will be formulated during the course of the evaluation. The listener should finally ask, "which of these relationships are significant to communicating the musical message, and which are not?"

When using artistic elements in the recording, the recordist should consider their relationships to the traditional musical elements and materials. This is necessary, to obtain an understanding of the importance of each musical material, as related to the piece as a whole. Through these observations, the recordist will obtain an understanding of the significance of the artistic elements, to communicating the message (or meaning) of the music.

This entire evaluation process will greatly assist the recordist in understanding how the artistic elements may be applied in the recording process to enhance, shape, or create musical materials and relationships.

Graphing the artistic elements may be time consuming and, at times, tedious. Graphing the activity of the various artistic elements is important for developing aural skills and evaluation skills, especially during beginning studies. It is also valuable for performing in-depth evaluations of recordings, providing insights into the artistic aspects of the recordist's own recordings and the recordings of others. This process of graphing the activity of the various artistic elements is also a useful documentation tool. Working professionals through beginning students will find the process useful, in a variety of applications.

Graphing the artistic elements is not proposed for regular use in professional production facilities and projects. It is not intended to be added to the production process itself. Audio professionals must be able to recognize and understand the concepts of the recording production and to hear many of the general relationships, quickly and without the aid of graphs. Graphs are intended to develop these skills and to provide a means for more detailed and in-depth evaluations that will take place outside of the production process itself.

Recordists who have developed a sophisticated auditory memory will also find these graphing systems of evaluation to be useful for notating their production ideas and for documenting recording production practices. These acts will allow them to remember and evaluate their production practices more effectively, allowing them more control of their craft.

The method for evaluating sound may be adapted for new concepts of recording production and new artistic elements. New technologies and artistic practices will lead to new artistic elements. A means to objectively evaluate those elements will be needed for these new concepts.

The processes and methods presented above are adaptable. How they may be adapted will depend on the new artistic elements. The method of evaluation must always conform to the artistic element being evaluated.

As to what the new artistic elements will be, we can only imagine and speculate, waiting with anticipation and listening with an open mind and an ever-shifting perspective, ever conscious that the unimaginable might occur at the next moment.

The future dimensions of *The Art of Recording* will be defined by its ever-evolving artistic elements, and how they are used for artistic expression.

Bibliography

Alten, Stanley R. 1990. *Audio in Media*, 3rd ed. Belmont, CA: Wadsworth Publishing Company.

Backus, John. 1977. *The Acoustical Foundations of Music*, 2nd ed. New York: W.W. Norton & Co., Inc.

Ballou, Glen. 1987. *Handbook for Sound Engineers: The New Audio Cyclopedia*. Indianapolis: Howard W. Sams & Company.

Bartlett, Bruce, and Michael Billingsley. 1990. An improved stereo microphone array using boundary technology: theoretical aspects. *Journal of the Audio Engineering Society* 38(7/8):543–552.

Bartlett, Bruce. 1987. *Introduction to Professional Recording Techniques*. Indianapolis: Howard W. Sams & Co., Inc.

Bartlett, Bruce. 1989. *Recording Demo Tapes at Home*. Indianapolis: Howard W. Sams & Co., Inc.

Bech, Søren, and O. Juhl Pedersen, editors. Proceedings of a Symposium on *Perception of Reproduced Sound*; Gammel Avernæs, Denmark, 1987. Peterborough, NH: Old Colony Sound Lab Books.

Beranek, Leo L. 1986, 1954. *Acoustics*. New York: American Institute of Physics, Inc.

Bergson, Henri. 1962. *Matter and Memory*. New York: Humanities Press.

Berry, Wallace. 1966. *Form in Music*. Englewood Cliffs, NJ: Prentice-Hall.

Blauert, Jens. 1969/70. Sound Localization of the Median Plane. *Acustica* 22:205–213.

Blauert, Jens. 1983. *Spatial Hearing*. Cambridge, MA: The MIT Press.

Blaukopf, Kurt. 1971. Space in Electronic Music. In *Music and Technology, Stockholm Meeting June 8–12, 1970*, pp 157–172. New York: Unipub.

Camras, Marvin. 1988. *Magnetic Recording Handbook*. New York: Van Nostrand Reinhold Company.

Chowning, John. 1977. The simulation of moving sound sources. *Computer Music Journal* 1(3):48–52.

Clifton, Thomas. 1983. *Music As Heard: A Study in Applied Phenomenology*. New Haven, CT: Yale University Press.

Cooper, Grosvenor W., and Leonard B. Meyer. 1960. *The Rhythmic Structure of Music*. Chicago: The University of Chicago Press.

Cooper, Paul. 1973. *Perspectives in Music Theory*. New York: Dodd, Mead & Company.

Davis, Don, and Carolyn Davis. 1975. *Sound System Engineering*. Indianapolis: Howard W. Sams & Co., Inc.

Davis, Don, and Chips Davis. 1980. The LEDE™ concept for the control of acoustic and psychoacoustic parameters in recording control rooms. *Journal of the Audio Engineering Society* 28(9):585–595.

Deutsch, Diana. 1982. *The Psychology of Music*. Orlando, FL: Academic Press, Inc.

Deutsch, Diana, and J. Anthony Deutsch. 1975. *Short-Term Memory*. New York: Academic Press.

Eargle, John. 1986. *Handbook of Recording Engineering*. New York: Van Nostrand Reinhold Company.

Eargle, John. 1981. *The Microphone Handbook*. Plainview, New York: Elar Publishing.

Eargle, John, editor. 1986. *An Anthology of Reprinted Articles on Stereophonic Techniques*. New York: Audio Engineering Society, Inc.

Erickson, Robert. 1975. *Sound Structure in Music*. Berkeley, CA: University of California Press.

Fay, Thomas. 1971. Perceived hierarchic structure in language and music. *Journal of Music Theory* 15(1–2):112–137.

Federkow, G., W. Buxton, and K. Smith. 1978. A computer-controlled sound distribution system for the performance of electronic music. *Computer Music Journal* 2(3):33–42.

Harris, John. 1974. *Psychoacoustics*. New York: The Bobbs-Merrill Company.

Hawking, Stephen W. 1988. *A Brief History of Time: From the Big Bang to Black Holes*. New York: Bantam Books.

Helmholtz, Hermann. 1967. *On The Sensations of Tone*. New York: Dover Publications, Inc.

Huber, David Miles. 1988. *Microphone Manual: Design and Applications*. Indianapolis: Howard W. Sams & Company.

James, William. 1950. *Principles of Psychology*. New York: Dover Publications, Inc.

Karkoschka, Erhard. 1971. Eine Hörpartitur elektronischer Musik. *Melos* 38(11):468–475.

Karkoschka, Erhard. 1976. *Neue Musik / Analyses*. Herrenberg: Doring.

Koffka, Kurt. 1963. *Principles of Gestalt Psychology*. New York: Harcourt, Brace, and World.

Kuttruff, Heinrich. 1979. Room Acoustics, 2nd ed. London: Applied Science Publishers Ltd.

LaRue, Jan. 1970. *Guidelines for Style Analysis*. New York: W.W. Norton & Company, Inc.

Leeper, Robert. 1951. Cognitive processes. In *Handbook of Experimental Psychology*, ed. S. S. Stevens, pp. 730–757. New York: John Wiley & Sons, Inc.

Letowski, Tomasz. 1985. Development of technical listening skills: timbre solfeggio. *Journal of the Audio Engineering Society* 33(4):240–244.

Martin, George. 1979. *All You Need Is Ears*. New York: St. Martin's Press.

McAdams, Stephen, and Albert Bregman. 1979. Hearing musical streams. *Computer Music Journal* 3(4):26–43.

Meyer, Leonard B. 1956. *Emotion and Meaning in Music*. Chicago: The University of Chicago Press.

Meyer, Leonard B. 1973. *Explaining Music: Essays and Explorations*. Berkeley, CA: University of California Press.

Meyer, Leonard B. 1967. *Music, the Arts and Ideas*. Chicago: The University of Chicago Press.

Miller, George. 1967. The magical number seven, plus or minus two. In *Language and Thought*, ed. Donald C. Hildum, pp 3–31. Princeton, NJ: Van Nostrand Company, Inc.

Mills, A. W. 1958. On the minimum audible angle. *Journal of the Acoustical Society of America* 30:237–246.

Moylan, William. 1983. *An Analytical System for Electronic Music*. Ann Arbor, MI: University Microfilms.

Moylan, William. 1985. Aural analysis of the characteristics of timbre. Paper presented at 79th Convention of the Audio Engineering Society, New York, NY.

Moylan, William. 1986. Aural analysis of the spatial relationships of sound sources as found in two-channel common practice. Paper presented at 81st Convention of the Audio Engineering Society, Los Angeles, CA.

Moylan, William. 1987. A systematic method for the aural analysis of sound sources in audio reproduction/reinforcement, communications, and musical contexts. Paper presented at 83rd Convention of the Audio Engineering Society, New York, NY.

Nisbett, Alec. 1979. *The Technique of the Sound Studio*, 4th ed. Boston: Focal Press.

Nisbett, Alec 1983. *The Use of Microphones*, 2nd ed. Boston: Focal Press.

Olson, Harry F. 1967. *Music, Physics and Engineering*, 2nd ed. New York: Dover Publications, Inc.

Pellegrino, Ronald. 1983. *The Electronic Arts of Sound and Light*. New York: Van Nostrand Reinhold Company.

Peus, Stephan. 1977. Microphones and transients. *db* 7:40–43.

Plomp, Reinier. 1976. *Aspects of Tone Sensation: A Psychophysical Study*. New York: Academic Press Inc.

Pousseur, Henri. 1959. Outline of a method. In *die Reihe, Nr. 3*, ed. Herbert Eimert and Karlheinz Stockhausen, pp. 44–88. Bryn Mawr, PA: Theodore Presser, Co.

Randall, J. K. 1967. Three lectures to scientists. *Perspectives of New Music* 3(2):124–140.

Reynolds, Roger. 1968. It(')s time. *Electronic Music Review* 7:12–17.

Reynolds, Roger. 1975. *Mind Models: New Forms of Musical Experience*. New York: Praeger Publishers.

Reynolds, Roger. 1978. Thoughts of sound movement and meaning. *Perspectives of New Music* 16(2):181–190.

Risset, Jean-Claude. 1978. *Musical Acoustics*. Paris: Centre George Pompidou Rapports IRCAM No. 8.

Roads, Curtis, editor. 1989. *The Music Machine*. Cambridge, MA: The MIT Press.

Roederer, Juan G. 1979. *Introduction to the Physics and Psychophysics of Music*, 2nd ed. New York: Springer-Verlag.

Rossing, Thomas D. 1990. *The Science of Sound*, 2nd Ed. Reading, MA: Addison Wesley Publishing Company.

Runstein, Robert E., and David Miles Huber. 1986. *Modern Recording Techniques*. Indianapolis: Howard W. Sams & Co., Inc.

Schaeffer, Pierre. 1952. *A la recherche d'une musique concrète*. Paris: Editions du Seuil.

Schaeffer, Pierre, and Guy Reibel. 1966. *Solfège de l'objet sonore*. Paris: Editions du Seuil.

Schaeffer, Pierre. 1966. *Traité des objets musicaux*. Paris: Editions du Seuil.

Schouten, J.F. 1968. The Perception of Timbre. *Report of the 6th International Congress on Acoustics*, 35–44, 90.

Smith, F. Joseph. 1979. *The Experiencing of Musical Sound: Prelude to a Phenomenology of Music*. New York: Gordon and Breach Science Publishers, Inc.

Stravinsky, Igor. 1970. *Poetics of Music: In the Form of Six Lessons*. Cambridge, MA: Harvard University Press.

Stevens, Stanley Smith, and Hallowell David. 1938, 1983. *Hearing: Its Psychology and Physiology*. New York: Acoustical Society of America.

Stevens, Stanley Smith, and E. B. Newman. 1936. The localization of actual sources of sound. *American Journal of Psychology* 48:297–306.

Stockhausen, Karlheinz. 1962. The concept of unity in electronic music. *Perspectives of New Music* 1(1):39–48.

Tenney, James. 1986. *META+HODOS and META Meta+HODOS*. Oakland, CA: Frog Peak Music.

Varèse, Edgard. 1966. The liberation of sound. *Perspectives of New Music* 5(1):11–19.

Wadhams, Wayne. 1990. *Sound Advice: The Musicians Guide to the Recording Studio*. New York: Schirmer Books.

Warren, Richard M. 1982. *Auditory Perception: A New Synthesis*. New York: Pergamon Press Inc.

Wertheimer, Max. 1938. Laws of organization in perceptual forms. In *A Source Book of Gestalt Psychology*, ed. Willis Ellis, 71–88. London: Routledge & Kegan Paul.

Wilson, David. 1984. Do you hear what I hear? *Mix Magazine* 8(6):132–134.

Winckel, Fritz. 1967. *Music, Sound and Sensation: A Modern Exposition*. New York: Dover Publications, Inc.

Winckel, Fritz. 1963. The psycho-acoustical analysis of structure as applied to electronic music. *Journal of Music Theory* 7(2):194–246.

Woram, John M. 1989. *Sound Recording Handbook*. Indianapolis: Howard W. Sams & Company.

Discography

The Beatles. "A Day In The Life." *Sgt. Pepper's Lonely Hearts Club Band.* CDP 7 464422. EMI Records Ltd., 1967, 1987.

The Beatles. "Every Little Thing." *Beatles for Sale.* CDP 7 46438 2. EMI Records Ltd., 1964.

The Beatles. "Lucy In The Sky With Diamonds." *Sgt. Pepper's Lonely Hearts Club Band.* CDP 7 46442 2. EMI Records Ltd., 1967, 1987.

The Beatles. "She Said She Said." *Revolver.* CDP 7 464412. EMI Records Ltd., 1966.

The Beatles. "Strawberry Fields Forever." *Magical Mystery Tour.* CDP 7 48062 2. EMI Records Ltd., 1967, 1987.

Bush, Kate. "This Woman's Work." *The Sensual World.* CK 44164. Columbia Records/CBS Records, Inc., 1989.

Chapman, Tracy. "Fast Car." *Tracy Chapman.* E2 60774. Elektra/Asylum Records, 1988.

Collins, Phil. "Hang In Long Enough." *...But Seriously.* 82050-2. Atlantic Recording Corporation, 1989.

Collins, Phil. "In The Air Tonight." *Face Value.* 16029-2. Atlantic Recording Corporation, 1981.

Gabriel, Peter. "Mercy Street." *So.* 9 24088-2. Geffen Records, 1986.

John, Elton. "Sorry Seems To Be The Hardest Word." *Elton John's Greatest Hits Volume II.* MCAD-37216. MCA Records, Inc., 1976.

The Police. "Every Breath You Take." *Every Breath You Take: The Singles.* CD 3902/DX 824. A & M Records, Inc, 1986.

U2. "Where The Streets Have No Name." *The Joshua Tree.* A2-90581. Island Records, Inc., 1987.

Yes. "Every Little Thing." *Yes.* 8243-2. Atlantic Recording Corporation, 1969.

Index